ECO-NATIONALISM

ECO-NATIONALISM

Anti-nuclear Activism and National Identity in Russia, Lithuania, and Ukraine

JANE I. DAWSON

Duke University Press Durham and London 1996

Printed in the United States of America on acid-free paper ∞
Designed by Cherie H. Westmoreland
Typeset in Sabon with Eras display type by Keystone Typesetting, Inc.
Library of Congress Cataloging-in-Publication Data appear
on the last printed page of this book.

For my parents

CONTENTS

PREFACE

This project spans a six year period, beginning in 1989 and continuing up until the present. At the time I embarked on this study, public activism in the Soviet Union was an entirely new and momentous phenomenon. After years of repression, the strictures on public speech and independent associations were breaking down, and ordinary citizens were slowly creeping out to test the waters of this newly tolerant system. After decades of silence, Soviet society was finally awakening. People began speaking out on controversial platforms and mobilizing to demand that the government open its ears to the concerns of its citizens. For those of us accustomed to the grim monotony of the Brezhnev era, Gorbachev's opening up of society was both welcome and exciting.

This project was undertaken in those heady days. With my own interests in environmentalism spurring me along, I embarked on an investigation of how people in the Soviet Union were utilizing these new opportunities to mobilize on environmental platforms. I soon discovered that the anti-nuclear cause was without contest the most potent mobilizing issue in the environmental realm—and perhaps even all realms—and I quickly dove into this issue as a way of looking at the novel phenomenon of public activism in the USSR.

My investigation, however, led me in unexpected directions. Rather than simply providing a window into the rebirth of civil society, my research revealed an unanticipated linkage between anti-nuclear activism and nationalism in many regions of the former USSR. I was startled to learn that rather than reflecting strongly held environmental principles, the movements against nuclear power were in fact often more indicative of popular demands for national sovereignty and regional self-determination. The shoddily constructed and carelessly operated nuclear power stations that littered the countryside were much more than environmental threats; they were in fact symbols of Moscow's indifference to the welfare and very survival of the non-Russian nations of the Soviet Union. Thus, what began as an investigation into the emergence of civil society in the late Soviet period evolved into a study of the fascinating phenomenon of eco-nationalism. In the chapters that follow, I ask how we can explain the convergence of environmentalism and nationalism in the late Soviet period and what the implications of this phenomenon are for the future development of environmental and anti-nuclear activism in the newly independent states of the former USSR.

A few words about methodology are in order. The detailed case studies which follow are based on extensive field research carried out in Lithuania, Ukraine, and Russia during the 1990–95 period. In order to gather the primary empirical information for the case studies presented, site visits were made to: Vilnius (Lithuania); Kiev, Rovno, Khmelnitsky, Chernobyl, Crimea (Ukraine); Moscow, St. Petersburg, Nizhny Novgorod, Rostov-na-Donu, and Tatarstan (Russia). At each site, the construction of the characteristics and developmental history of the local anti-nuclear power movement was based on a combination of primary sources, including in-depth interviews, local press reports, and official documents.

For each case, interviews were conducted with a variety of movement participants and observers. Leading members of the region's anti-nuclear and environmental organizations, journalists and scientists working in this issue-area, members of the environmental committees of city and regional councils, and representatives of the local branch of the State Committee for the Protection of Nature were interviewed to provide a view into environmental and anti-nuclear activities in the region. Proponents of nuclear power, including local nuclear specialists as well as representatives of the local nuclear power station administration were also interviewed to provide an alternative view of the anti-nuclear movement.

In order to understand how the local anti-nuclear movement fit into the spectrum of informal activities in a particular region, representatives of the leading political associations were included in the interview process. Not only did these interviews provide insight into the linkages between anti-nuclear and political associations, but they also provided a highly informed view of anti-nuclear activities in the region that often avoided the insider subjectivity of the anti-nuclear activists themselves.

In putting together a detailed picture of local anti-nuclear movements and their development, every attempt was made to confirm the information provided by movement participants and observers with printed, documentary evidence. The central and local press provided a rich source of information on movement activities, which was extremely useful in supporting and clarifying information gleaned from interviews as well as filling in gaps. In addition, most local anti-nuclear organizations kept archives containing copies of letters of protest, petitions, local press reports, and copies of key local decisions affecting the nuclear power station. Activists in all regions were extremely generous in providing me with complete access to these archives and frequently with copies of all key documents. Finally, members of the environmental committees of local soviets were also often very forthcoming in providing documentation of local decisions made on this issue.

I am enormously indebted to numerous sociologists, activists, nuclear specialists, and government officials in Lithuania, Ukraine, and Russia for invaluable assistance on this project. In particular, sociologists Olga Lobach, formerly of the Institute of Nuclear Safety in Moscow, and Olga Tsepilova, of the Institute of Sociology in St. Petersburg, went far beyond the call of duty in providing information and insight on anti-nuclear activities and assisting me in my efforts to travel to regions of the former USSR previously unexplored by Western scholars. In Moscow, economist and nuclear specialist Yurii Koryakin was tireless in his effort to provide me with the best possible information and contacts in the nuclear industry, and to keep me informed of the governmental decisions in the realm of nuclear power. While numerous environmental activists were generous in granting interviews and providing information on their movements, a few stand out for their exceptional generosity: Andrei Glazovoi, Vitalii Kononov, and Andrei Demidenko of the Ukrainian group Zelenii svit were particularly gracious in patiently fielding my questions year after year. To the many people in Lithuania, Ukraine, and Russia who made this project possible, I am sincerely grateful.

I would also like to thank the many scholars who provided invaluable comments and criticisms on this project from its inception to its completion. I am particularly grateful to my graduate advisors, Gail Lapidus,

George Breslauer, Ken Jowitt, and John Holdren, who unstintingly shared their knowledge and experience with me and were a constant source of new ideas and inspiration. I am also indebted to my colleagues in the Berkeley-Stanford Program in Soviet Studies who provided constant feedback, ideas, and support throughout my years of graduate study and after. Discussions with Kelly Smith, Kathie Hendley, Shari Cohen, Jason McDonald, Matt Trail, and Clay Moltz played a significant role in shaping this project. Numerous other scholars have been generous with their time and insights, and in particular I would like to thank Tim Colton, David Holloway, Craig Murphy, Bill Potter, Susan Solomon, and Steve Weber for useful comments on this and earlier drafts. I am also grateful to my colleagues at Wellesley College, where I put the finishing touches on this project, for their lively interest, continuous support, and many useful insights. Finally, I would like to express my immense gratitude to my husband and colleague, Rob Darst, whose moral and intellectual support was critical in bringing this project to fruition.

I would also like to acknowledge the generous support of several foundations, which made both the dissertation research and later follow-up research for this project possible. Assistance from the Joint Committee on Soviet Studies of the Social Science Research Council, the Institute for the Study of World Politics, the Institute on Global Conflict and Cooperation, the National Council for Soviet and East European Research, and Wellesley College was invaluable to the completion of this project.

ECO-NATIONALISM

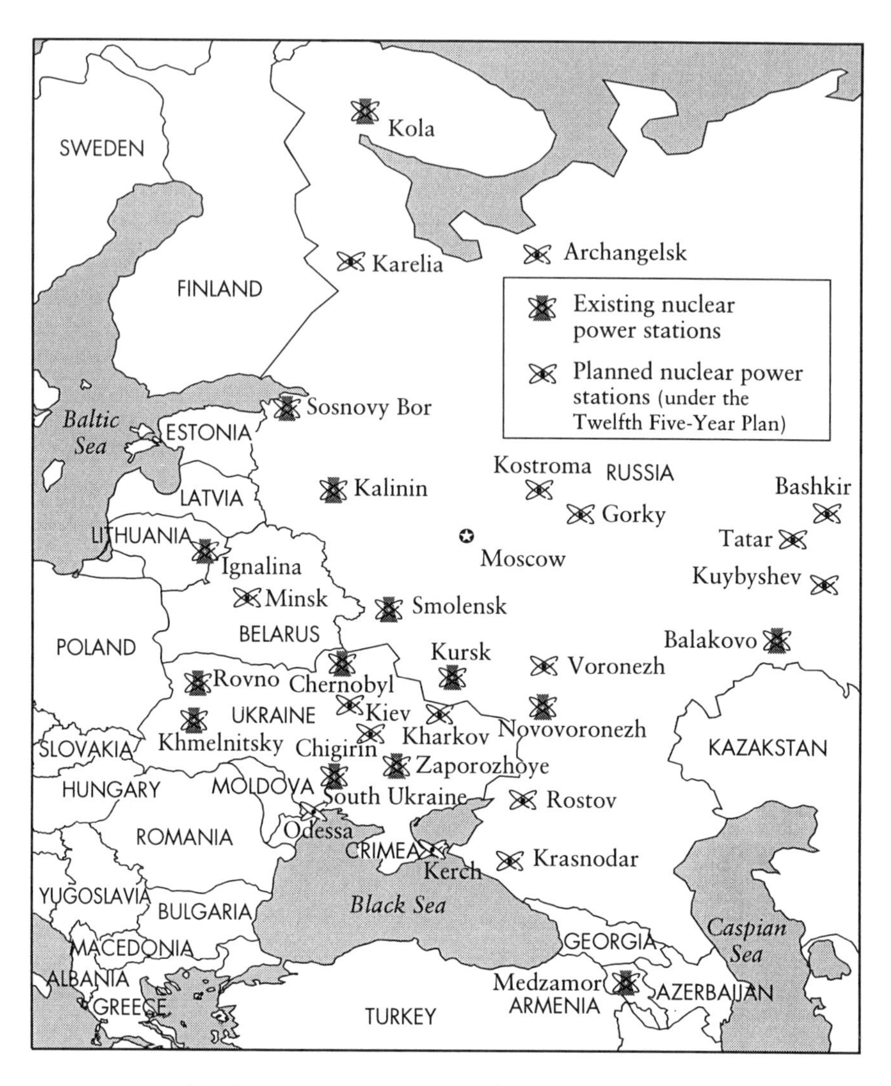

Nuclear Power Stations in the Western USSR

INTRODUCTION

Anti-nuclear Activism in Comparative Perspective

The story of nuclear power is the story of a dream gone awry. What began as an ambitious and promising effort to alleviate dependence on limited supplies of fossil fuels has come to be seen as one of the greatest technological follies in modern history.[1] Around the globe, popular fears and distrust have repeatedly derailed ambitious programs for the expansion of nuclear power. In country after country, plans to bring new nuclear reactors on line have been stymied, and in some cases public opposition has even forced the closure of stations already in operation. Clearly, the early proponents of nuclear power overlooked a potent factor when proposing the construction of immense power stations relying on this new technology. Over the past two decades, citizen opposition to nuclear power has emerged as a critical element in shaping nuclear power decisions around the globe.[2]

While numerous countries, including the United States, Britain, Germany, France, and Japan, adopted ambitious programs to dramatically expand their reliance on nuclear power in the early 1970s, all of these governments found themselves confronted with substantial, and often vociferous, popular opposition to their chosen path a decade later. Although a few of these countries, particularly France and Japan, were able to insulate nuclear decision making from public opposition and thus permit the continuation of their expansion programs,[3] most of these democracies were overwhelmed by an explosion of popular fear and mistrust. In the United States and across much of Western Europe, citizen mobilization proved highly successful in forcing government and industry to abandon plans for the expansion of nuclear power.

While the advanced industrialized democracies found their nuclear power programs under fire in the late 1970s and early 1980s, the communist world appeared strangely calm. While the Soviet Union had also adopted an ambitious program to dramatically expand nuclear power production in the USSR and Eastern Europe in the early 1970s, a decade later no signs of popular opposition to nuclear power could be discerned in the region. The communist systems of the Soviet Union and Eastern Europe appeared entirely immune to the anti-nuclear disease, which had spread so quickly throughout the West. In fact these governments appeared able to proceed with their ambitious nuclear plans with absolute impunity.

The absence of anti-nuclear opposition and its consequent impact on decision making in the USSR (as well as Eastern Europe) is readily understandable when one considers the nature of the Soviet political system prior to 1985. As a system based on the Communist Party's monopolization of the public realm, the Soviet model functioned to prevent the coalescence of independent and unsanctioned public activities.[4] Opportunities for opponents of nuclear power to speak out publicly, to appeal to mass audiences, and to organize resistance to the state's official commitment to the expansion of nuclear power in the USSR were almost nonexistent prior to 1985. Furthermore, the Party's control over the dissemination of information provided the state with a powerful tool to prevent the spread of anti-nuclear opposition in Soviet society. The state's ability to block the flow of information from outside the USSR's borders while simultaneously propagandizing its own view on the desirability of nuclear power, ensured that the public would remain positively disposed toward the state's energy agenda. Thus the anti-nuclear contagion was effectively halted at the borders of the USSR.

All of this changed, however, with the watershed year of 1986. With the introduction of Mikhail Gorbachev's reform program and the graphic demonstration of the dangers of nuclear power provided by the horrific accident at the Chernobyl nuclear power station on April 26, 1986, opportunity and impetus for anti-nuclear mobilization suddenly converged in the Soviet Union.[5] More than a decade after opposition to nuclear power had begun to sweep through the United States and Western Europe, the Soviet public slowly began to awaken to the dangers of this powerful technology. As the public became ever more aware of the devastating consequences of the Chernobyl disaster and as opportunities for mass mobilization expanded, popular opposition to nuclear power gathered momentum and soon became a potent force in shaping the Soviet government's nuclear power program.

From 1987 to 1991, popular movements opposing the continued operation or construction of nuclear reactors proliferated rapidly across the four republics which were home to the Soviet Union's nuclear power program: Armenia, Lithuania, Ukraine, and Russia. This explosion of anti-nuclear activism eventually encompassed virtually every nuclear power station in operation or under construction in the USSR.[6] After unquestioning acceptance of the government's ambitious plans to double nuclear output during the Twelfth Five-Year Plan period (1986–90), the Soviet population suddenly rose up to demand that the government cancel plans for all future reactors, halt construction projects, and close entire stations.

From 1988 to 1990, anti-nuclear activists were unexpectedly successful in pressuring officials to accede to movement demands. In Lithuania, plans to expand the republic's sole nuclear power facility were canceled,[7] while in Armenia the only operating nuclear power station was closed completely.[8] In Russia and Ukraine, plans for one new project after another fell in the face of overwhelming popular opposition, and construction projects already under way were suspended or occasionally halted completely. In all, the construction of over forty nuclear reactors was either canceled or suspended during this period.[9] Eventually, the parliaments of both Russia[10] and Ukraine[11] passed five-year moratoriums on the construction of new nuclear facilities in their territories and thus brought the Soviet Union's nuclear power program to a virtual standstill.

Despite the apparent success of these movements, however, they eventually failed to meet their larger objectives. Rather than forming the basis for a powerful anti-nuclear movement, which would fight for the closure of the more than forty nuclear reactors still operating on Soviet soil, public activism on this issue proved unexpectedly short-lived. The movements burst forth in the early period of perestroika, then faded away, leaving little evidence of their previous existence. While some of the movements were successful in preventing the expansion of nuclear power in their regions, little progress was made in ridding the territory of the Soviet Union of the perils of nuclear reactors. Despite the horrific experience of the Chernobyl accident and its graphic demonstration of the potential dangers of Soviet-constructed nuclear power stations, interest in the nuclear power issue withered away, and by 1991, few signs of the previous strength and dynamism of Soviet anti-nuclear power movements remained. In 1991, the Lithuanian parliament renewed discussions on expanding the Ignalina nuclear power station, while in the Armenian republic, officials began to explore the possibility of reopening the closed Medzamor station. By late 1993, the governments of both Russia and Ukraine had annulled their moratorium on the construction of new facilities[12] and had begun to take steps to resume numerous previously suspended or canceled nuclear projects. And by 1995, the Armenians were actively preparing to reopen their sole nuclear power station (scheduled to go back on line in May of 1995),[13] while new reactors had already been brought into operation in Ukraine. Little public outcry was heard in response to these reversals.

Eco-nationalism: *The Convergence of Environmentalism and Nationalism*

Paradoxically, while anti-nuclear power movements failed to create the foundation for a unified and lasting anti-nuclear or environmental movement in the USSR during the perestroika period, they proved powerful springboards for the creation of national movements in the non-Russian republics and regions of the Soviet Union.[14] In both Lithuania and Armenia, anti-nuclear movements carried strong nationalist overtones from their inception and were rapidly transformed and incorporated into movements for republic sovereignty. In Ukraine, where national identity was buried beneath decades and even centuries of Russification, anti-nuclear movements provided a context through which national consciousness could be forged and mobilizational networks and skills created. And in Tatarstan, an "Autonomous Soviet Socialist Republic" within the Russian Federation, the anti-nuclear movement represented the first step toward the revival of the long-dormant Tatar nation.

In contrast, in Russian regions of the Russian Federation, activists were unable to identify a national group to juxtapose against the imperial center, and movements were limited to the single issue of nuclear power.

While Russian movements attempted to take advantage of demands for greater territorial self-determination, they were unable to find a strong group identity to mobilize and thus suffered substantial mobilizational limitations relative to their counterparts in the non-Russian republics. Understanding this changing linkage between anti-nuclear activism and nationalism in different regions and republics of the former USSR and its successor states is one of the main objectives of this study.

This is the story of the rise and fall of the anti-nuclear power movement in the USSR and its successor states.[15] The evolution of this movement, however, is inextricably linked to the phenomenon of eco-nationalism. Thus we must ask: What was it about the political, social, and economic conditions of the late Soviet period that favored the convergence of environmental and national objectives? Why was the anti-nuclear cause adopted as the centerpiece of the nationalist movements of Armenia, Lithuania, and, to a lesser extent, Ukraine but not effectively linked to nationalism in Russia? What role does national identity play in determining whether environmentalism will take on a nationalist hue? And finally, what is the significance of the eco-nationalist phenomenon for understanding environmental activism in other parts of the world? Did an unusual convergence of conditions and identities create a unique phenomenon in the Soviet case or might environmentalism and nationalism go hand in hand in other parts of the world?

Theoretical Tools: *Considering Resources and Identities*

In comparing mobilizational patterns in the former USSR and the advanced industrialized world of the West, differences in the characteristics and developmental trajectories of anti-nuclear movements are readily apparent. In this study, I argue that structural factors provide a good starting point for explaining the unexpected mobilizational patterns observed in the USSR and its successor states. The Communist Party's longstanding monopolization of mobilizational resources left an important structural legacy, which has dramatically shaped the ability of autonomous actors to mobilize independent movements.[16] In the following chapter, I suggest that the most significant aspects of this legacy include: (1) the scarcity of mobilizational resources in society and (2) uneven and constantly changing access to key resources by competing independent actors.

This dual legacy dramatically affected the characteristics of the social movements that emerged in the former USSR and its successor states after 1985. First, the scarcity of mobilizational resources for actors outside the Communist Party substantially undermined the possibility for the creation of tightly organized social movements during the perestroika period (1986–91). During this period, limited access to critical resources led movement organizations to rely largely on indigenous leadership and volunteer staff (rather than the professionalized movements common in the developed West),[17] weak linkages both within and between movement branches, unstable group membership, and disruptive tactics. The continuing dearth of key mobilizational resources during the transitional period following the collapse of the USSR in 1991 has made these organizational limitations a persistent feature of post-Soviet societies. Indigenous movements continue to be haunted by shortages of critical tangible resources, such as funds, meeting space, and communications technology, as well as intangibles, including organizational skills and networks for bloc recruitment. Interestingly enough, with the flood of foreign capital into the region after 1991, the most vibrant and effectively organized movements of the transition period are frequently those directly linked to and sponsored by wealthy foreign and international organizations.

The second aspect of the communist legacy that I focus on in this study is the ever changing distribution of mobilizational resources among competing groups in society. This also significantly shaped mobilization patterns throughout both the perestroika and transitional periods. During the perestroika years, the Communist Party's ad hoc dissemination of critical resources to groups outside the Party opened opportunities for some movements to thrive while simultaneously blocking possibilities for other movements to emerge. Thus, during the early perestroika years, movements deemed nonthreatening to the Party's status and goals were privileged in their access to important resources, while independent actors with more radical agendas were denied opportunities and resources. This created the conditions under which *movement surrogacy* might emerge. During the initial period of perestroika, it was not uncommon for radical actors to hide behind surrogate causes that targeted similar audiences. In the upcoming chapters, I will argue that in some parts of the former USSR, the anti-nuclear movement was little more than a surrogate for hidden nationalist demands.

As the mobilizational environment changed, however, access to key resources shifted to more radical actors and the need for movement surrogacy disappeared. Surrogacy was simply a rationally selected tactic that fit the existing structural conditions and that was discarded when no longer useful. Thus, the convergence of environmentalism and nationalism during the perestroika period may be at least partially accounted for by the unusual structural conditions under which these movements emerged.

Since 1991, the distribution of mobilizational resources in society has continued to shape the development of social activism in the newly independent countries. While the distribution of mobilizational resources in society no longer encourages the intertwining of environmental and nationalist goals, it nonetheless continues to affect movement characteristics in other ways. As noted above, one of the most significant aspects of the post-Soviet period has been the unusually privileged position of foreign-sponsored organizations in the Soviet successor states. Since 1991, Western aid has become a critical factor in determining which independent organizations will survive and be able to carry out their agendas. This new factor has led to a substantial change in mobilizational patterns since 1991.

While a structural analysis offers significant insight into mobilization in late and post-Soviet society, it does not provide a complete picture of the mobilizational process. In this study, I also consider the impact of identity in shaping how people mobilize. I argue that the preservation and possible reinforcement of national identities during the communist period (and absence of strong alternative group identities) is a critical factor in explaining the linkage between nationalism and anti-nuclear activism during the perestroika years. The anti-nuclear movements of the former USSR were more than simple protests against a potentially dangerous technology; they were the cries of colonized nations against the antidemocratic incursions of an imperial center. It is this linkage between anti-nuclear activism and national identity that most significantly differentiates the anti-nuclear movements of the former Soviet Union with their counterparts in the West.[18] Because the strength and the characteristics of national identities differ across Russia, Ukraine, Lithuania, and Armenia, however, the ways in which anti-nuclear and nationalist causes became intertwined varied by region. In the case studies that follow we will probe this variation and consider the role of national identity in shaping the potential for eco-nationalism to emerge.

With the achievement of national independence for the fifteen republics of the former Soviet Union in 1991, the role of national identity in mobilizing anti-nuclear protest diminished substantially. While minority nationalities could still rail against the domination of foreign overlords, the titular nationalities now found themselves masters of their own fate. No longer could they equate environmental contamination with colonial oppression. No longer did the local nuclear power station represent Moscow's careless treatment of the periphery; instead, the station came to be seen as a critical factor in the economic survival of the newly independent states. It represented self-sufficiency rather than imperial domination, and as such aroused little popular opposition. With independence, the symbolic function of the anti-nuclear movement evaporated, and popular interest in the issue plummeted. Since 1991, apathy has replaced environmental and anti-nuclear activism. With the linkage between environmentalism and nationalism effectively severed, the movement has fallen on hard times across all of the former USSR.

Overview

This theoretical synthesis of structural and identity models and its application to the late- and post-Soviet case is presented in detail in chapter 1. The empirical core of this study, however, is contained in the following five chapters. In these chapters, the anti-nuclear movements of Lithuania, Russia, Ukraine, and the national enclaves of Tatarstan and Crimea are thoroughly investigated.[19] Comparison of the movements in these different regions and now-independent states reveals both important similarities and differences in characteristics and development. Most important among these distinctions is the varying relationship between anti-nuclear activism and national identity. In the final chapter, I return to the theoretical framework presented in chapter 1 and consider how it may be refined to account for this changing linkage between anti-nuclear activism and nationalism.

Finally, I close with some thoughts on how anti-nuclear activism in the newly independent states may evolve in the future and what the consequent implications for the development of nuclear power in the Soviet successor states are. Given widespread Western concerns about the safety of the dozens of nuclear power stations now operating in Russia, Ukraine, Lithuania, and Armenia, the role of society in acting as a watchdog over the nuclear power sector in the Soviet successor states may be seen as particularly important. Plummeting popular interest in the nuclear power issue and evidence that the surge of anti-nuclear sentiment during the perestroika period was more politically than environmentally motivated, however, bode poorly for the likely role of post-Soviet society in watching over their nuclear power industry. The implications of these conclusions for ensuring the safe operation of nuclear power in the Soviet successor states are disturbing.

CHAPTER 1

Patterns of Social Mobilization in Late- and Postcommunist Societies

Since the explosion of popular protest in the West in the 1960s, considerable scholarly attention has been devoted to expanding our understanding of how people mobilize. Under what circumstances do people join together and publicly express common concerns and demands? Why do patterns of mobilization differ from group to group and across societies and political systems? In the past several decades, two important paradigms have emerged which provide insight into these questions. Resource mobilization theory, a structural approach, has focused attention on the role of access to mobilizational resources and opportunities in shaping patterns of mobilization.[1] In contrast, an identity-oriented model points to the role of ideas, culture, and group identity in explaining how mobilization occurs.[2] While initially viewed as competing models, it has more recently been suggested that these two paradigms may be utilized in tandem to provide a richer and more complex view into the dynamics of popular mobilization.[3]

While these two models were developed in response to the explosion of social activism in Western societies in the 1960s and beyond, they provide an excellent jumping-off point for interpreting and comparing social-movement dynamics in other types of societies, particularly communist and postcommunist. In this chapter, I will (1) briefly review the basic tenets of the resource-mobilization and identity-oriented approaches and their synthesis; (2) employ a hybrid framework drawn from these two approaches to hypothesize how people might be expected to mobilize in late- and postcommunist societies; (3) consider the patterns of antinuclear mobilization observed in Lithuania, Russia, and Ukraine from

1985 to 1995 and provide a preliminary assessment of the utility of this Western paradigm in the study of social movements in late- and postcommunist societies.

Determinants of Social Mobilization: *Resources and Identity*

While the dominant approach to collective behavior of the 1950s and 1960s focused attention on the role of individual grievances in triggering mass mobilizations, the new resource mobilization theory that emerged in the 1970s firmly rejected this focus on grievances, deprivation, and anomie.[4] Rather than focusing on why individuals mobilize, resource mobilization theorists shifted their attention to how collective actors pursue their goals within a given structural context. Within this perspective, grievances are no longer viewed as primary determinants of mobilization. Instead, structural factors such as resources, organization, and opportunity become the key determinants of when and how collective behavior may occur. Grievances in society are considered to be sufficiently bountiful that they are always available for mobilization; movements only occur, however, when structural changes facilitate their emergence. In its most extreme variant, McCarthy and Zald have argued that "the definition of grievances will expand to meet the funds and support personnel available."[5]

Resource mobilization theorists focus on social movements as rational actors operating on the basis of cost-benefit analyses. Collective actors are assumed to utilize strategic instrumental rationality to determine how best to pursue movement goals within a given context. Given specific resource availability, preexisting organizational form, and opportunities, the collectivity will rationally select its tactics and strategy to maximize its potential for success. Thus, according to this school of thought, knowledge of resource availability, of organization, and of opportunity structures should yield a greater understanding of mobilizational patterns utilized by a particular social movement. Movement tactics, development, and level of success are expected to depend primarily on these structural factors.

While the resource mobilization approach has proven very useful for interpreting the activities of collectivities, it fails to address the important question of why individuals choose to participate in social movements.

While collectivities are assumed to base their actions on strategic calculations, the rational-actor assumption runs into Mancur Olson's collective-action problem when applied to individual movement participants. According to Olson, in the absence of coercion or selective benefits, it is more rational for individuals to free ride than to participate in collective action.[6] How then can we explain the existence of countless social movements?

An answer to this question was suggested by proponents of an alternative identity-oriented theory of social movements. Focusing primarily on the "new social movements" characteristic of postindustrial societies, theorists such as Pizzorno, Melucci, and Touraine pointed to the expressive function of social movements. Rather than viewing movements as rational actors, these theorists saw social movements as fora in which people could express feelings and search for meaning and identity. Movements thus provide a mechanism through which social identities are shared and negotiated. The key to mobilization is not objective but rather subjective factors.

While resource mobilization theorists were initially reluctant to cloud their crisp, objective model with elements of subjective motivation, over the past several years a consensus regarding the necessity of integrating the two approaches[7] has been growing. The inability of resource mobilization theory to adequately address the question of why people participate has led to a new acceptance of the importance of expressive motivation in the creation and development of social movements.[8] This integration of objective and subjective motivational factors is not new; sociological luminaries including Weber and Habermas have long maintained the importance of both material and ideal interests in motivating action. It is reasonable to believe that the exclusion of one or the other, while vastly simplifying theory, is unlikely to provide a complete picture of human action.

The resource-identity hybrid that is currently emerging attempts to balance the role of structural factors in determining movement profiles and development with the expressive function of movements. It is thus acknowledged that individual participation in movements is better explained in subjective rather than objective terms. Participants are thought to be seeking a forum through which to express their feelings and to strengthen self-identity. In charting the development of a movement, it is therefore necessary to look beyond strategic calculations and to also consider the movement's function in affirming and developing group identities.

Resources and Identity in Communist and Postcommunist Societies

While the resource mobilization and identity-oriented paradigms were developed to explain mobilizational processes in the capitalist democracies of the United States and Western Europe, they also offer a promising lens through which to view mobilization in communist and postcommunist systems. In focusing on the availability of mobilizational resources, organizational forms, and opportunities, resource mobilization theory provides a powerful explanation for the absence of independent collective activities in the USSR prior to 1985 and the emergence of social activism during the perestroika years. Furthermore, by adding the identity element to the framework, this perspective draws attention to the distinctive identities created and nurtured by the communist experience. Thus, the resource and identity hybrid appears to be a particularly promising tool for investigating mobilizational patterns in both communist and postcommunist systems.

Prior to the introduction of Mikhail Gorbachev's reform program in the USSR, the Communist Party's monopoly control over the public realm was widely acknowledged to be a defining feature of the Soviet system. While scholars frequently disagreed over how best to conceptualize Soviet state-society relations, most agreed that the absence of opportunities for independent public activities represented an important cornerstone of the system.[9]

From the perspective of the resource mobilization framework, the Communist Party's monopoly control over the public realm easily translates into party monopolization of access to mobilizational resources. More specifically, communist systems may be defined by the party's control over both tangible and intangible mobilizational resources.[10] Thus, important tangible resources such as funds, meeting space, and communications technology are all tightly held within the party's grip. Similarly, independent actors are severely disadvantaged in access to less-tangible resources such as social networks, organizational skills, specialist expertise, and legal protection for their activities. A key goal of the Communist Party is to restrict public access to mobilizational resources and eliminate opportunities for independent groups to coalesce and express their interests publicly.[11] Conversely, the party itself is expected to act as a mobilizational machine which can quickly and efficiently mobilize its cadres in support of party interests.[12]

With the introduction of Gorbachev's program of perestroika in 1985, this core characteristic of the communist system began to unravel. By introducing glasnost into the Soviet system, Gorbachev was signaling that the party would no longer maintain monopoly control over the public realm. Instead, opportunities for individuals to publicly voice their own concerns and demands were to be expanded. In addition, Gorbachev moved quickly to encourage people to join together and form independent associations to discuss their concerns and ideas. All of this was part of Gorbachev's strategy of energizing the "human factor" to achieve the political and economic revitalization of Soviet society that was so desperately needed.

In opening up the public realm to independent actors and collectivities, Gorbachev initiated the process of shifting both tangible and intangible mobilizational resources from the party to other competing entities in society. The process through which the party relinquished its privileged access to mobilizational resources, however, was neither smooth nor regularized. During the early years of perestroika, this shift in resources was accomplished solely through Gorbachev's exhortations to the party and to society.[13] Despite Gorbachev's constant encouragement of new public activities, however, independent actors suffered from inadequate resources and opportunities for mobilization. New laws guaranteeing freedom of speech and association were not even put in place until 1990.[14] This uneven process through which resources were gradually distributed to independent groups played an important role in shaping the new social movements that began to emerge in the late 1980s.

The communist experience left a lasting legacy for independent groups attempting to mobilize during the Gorbachev period and after. Three aspects of this legacy are particularly important for understanding how people in the USSR and its successor states have mobilized since 1985. First, the party's longstanding monopolization of resources combined with the economic deterioration of the country led to a general scarcity of mobilizational resources for these newly emerging independent actors. Second, the ad hoc transferal of mobilizational resources from party to society led to an uneven distribution of these resources among competing groups; some groups were privileged in their access to resources, and this access changed constantly over time. Finally, the communist experience greatly shaped the types of group identities that were available for mobilization in the late 1980s and beyond. This threefold legacy of resource scarcity, of uneven and evolving distribution of mobilizational resources, and of identities emerging from the communist experience has played a powerful role in shaping mobilizational patterns both during the perestroika period and later. In order to understand how the communist legacy has affected social movement characteristics and development, we need to take a closer look at each aspect of this legacy. In the following sections, we will first examine the impact of this triple legacy during the perestroika years and then turn our attention to the post-1991 transition period.

Resource Scarcity

During the seventy years of communist rule in the USSR, significant progress was made in leveling inequalities and creating the basis for an egalitarian society in the Soviet Union.[15] Despite moderate success in meeting this primary ideological objective, the regime failed in its goal of sustaining rapid economic growth and overtaking the capitalist world in economic efficiency and productivity.[16] The result of the communist experiment was a society in which all (outside the nomenklatura) were relatively equal, but in deprivation rather than prosperity. The low average standard of living which characterized Soviet society in the late 1980s and has continued to haunt most of the fifteen successor states indicates that resource mobilization studies focusing on how resource-poor constituencies mobilize in Western societies are likely to provide useful insight into Soviet and post-Soviet social movement dynamics.[17]

In addition to a general lack of material resources in Soviet society, more subtle disadvantages in access to important tangible and intangible mobilizational resources for nonparty activities are expected to play a critical role in determining movement organization and tactics. In addition to limited access to funds, the newly emerging independent actors of the late 1980s found themselves seriously disadvantaged in comparison with the Communist Party with respect to other tangible mobilizational resources, such as access to meeting space and communications technology, and intangible resources, including organizational skills, specialist expertise, and legal protection. Low access to key mobilizational resources by these newly created collectivities lead us to expect (1) classical social movement organization based on grass roots, with indigenous leadership and volunteer staff; (2) weak vertical and horizontal linkages between units; (3) weak intragroup linkages; (4) unstable group membership; and (5) preference for disruptive tactics.

While professional "social movement organizations" (SMOs), with paid staff and outside leadership, are expected to predominate in contemporary Western societies, their formation and maintenance depend heavily on the ability of the movement to secure substantial financial contributions from individuals outside the aggrieved group.[18] In a society in which individuals had long been denied ownership of land and capital, and the overall standard of living was low, opportunities to accumulate significant resources from individual donors, however, were small. While donations from official institutions were possible in the USSR, they were subject to limitations in their quantity and distribution. Thus, Soviet movements would be far more likely to conform to the pattern of classical SMOs: relying on the time and energy of indigenous leadership, volunteer staff, and mass membership rather than material resources.

Due to the shortage of material resources, organizational objectives were likely to be limited by difficulties in obtaining permanent office and meeting spaces and by lack of adequate organizational and communications technology. Computers, copying machines, printing facilities, telefaxes, telephones, and transportation were likely to be in short supply, thus severely limiting the ability of movement members to organize and coordinate their activities. Loose organizational structures and weak linkages between territorially dispersed chapters would be expected. Given these weaknesses in communications capabilities, it is likely that linkages between movement leaders and followers would be weak. It is thus expected that a small core of devoted leaders might direct the activities of local chapters without regular or substantial input from the membership at large.

Membership is likely to be highly unstable outside the inner core due to lack of resources for maintaining membership networks through mass mailings, weekly meetings, and other regular communications. Stable participation in social movements is also likely to be constrained by growing economic shortages, which require consumers to spend more time in the search for goods and severely limit the time they spend on nonessential activities. The magnitude of the perceived threat to group interests is thus likely to be an important factor in determining the number of movement participants; in the absence of an overwhelming and tangible threat, people are unlikely to find the time to engage in collective activities.

The tactics utilized by newly mobilized groups are also expected to be constrained by resource scarcity. Costly tactics, such as employing professional lobbyists, advertising extensively in the mass media, and providing substantial support for desired election candidates, are likely to be beyond the financial resources of most emerging movements in the Soviet Union. Thus, tactics requiring little investment will be favored. Disruptive tactics, such as mass demonstrations, strikes, civil disobedience, and violence, have predominated among poor constituencies in Western societies.[19]

In sum, the newly created associations and movements of the perestroika period are expected to bear little resemblance to their contemporary Western counterparts. Rather than fitting the professional social movement model of resource-rich constituencies and societies, the new movements are more likely to display the characteristics of classical social movements of an earlier era in the West and of resource poor constituencies.

Uneven Distribution of Mobilizational Resources

In considering the impact of resource scarcity on mobilizational patterns in late- and postcommunist societies, a very important factor has thus far been ignored. In capitalist societies, distribution and access to goods is regulated by an impersonal market; this is not true in communist societies. Instead of private ownership and impersonal exchange in the market, ownership and distribution of land and capital in communist societies are the domain of the state. In addition, the Communist Party/state also monopolizes intangible resources, such as networks for bloc recruitment, and organizational skills. Thus, access to resources required for mobilization is strictly dependent on state preferences. The dissemination of mobilizational resources to competing groups in late-communist societies is therefore a highly political process.

In the former Soviet Union, Mikhail Gorbachev's insistence on utilizing the human factor to revitalize the ailing system led directly to a shift in resources from party to society. As noted earlier, however, this redistribution of mobilizational resources occurred on a very ad hoc basis, pushed ahead solely through Gorbachev's exhortations, and was not even cemented in law until 1990. Thus, the party organization relinquished its monopoly control over the public realm only fitfully and largely reluctantly throughout the perestroika period. Party preferences would continue to be an important factor in determining the winners and losers in the mobilizational process throughout much of the perestroika period.

Because the party played such an important role in controlling the dissemination of mobilizational resources during the perestroika period, it is reasonable to expect some groups to be privileged over others in their access to important resources. Groups with platforms deemed unthreatening to the party's status and goals were likely to be viewed more benignly than those representing a challenge to the party. Over time, however, as the party's always-dubious legitimacy collapsed and its control over public activities eroded, the role of party preferences in shaping the mobilizational environment would likely diminish. Those groups initially privileged by party preferences would not be expected to be able to maintain their advantages indefinitely. With the collapse of the party's role these privileged groups were likely to face stiff competition from newly emerging organizations and movements.

During the initial period, however, when groups with nonthreatening platforms experienced substantial advantages relative to their more radical counterparts, the phenomenon of *movement surrogacy*—the hiding of political intentions behind an apparently nonthreatening cause in order to take the first steps toward mobilizing people to support more radical platforms—could emerge. If both the explicit and implicit goals of the leading activists targeted the same audience, then this surrogate strategy might be viewed as the most rational strategy given the existing structural conditions. As new opportunities for more radical collective action emerged, however, this strategy could be discarded in favor of more open support for the activists' true agenda. Thus, the phenomenon of surrogacy would likely be short-lived, given the rapidly changing mobilizational environment of the perestroika period.

In addition to affecting movement strategies, the constantly changing distribution of mobilizational resources in society can also be expected to shape choice of tactics. While the party's immense superiority in mobilizational resources during the early perestroika period would leave independent actors with few choices other than petitions, popular protests, and other disruptive actions, the emergence of new opportunities for influence is likely to cause independent actors to shift their tactics. Particularly significant would be the introduction of quasi-democratic elections in 1989 and 1990 and the concurrent shift in power from the party to the newly elected organs. With this change, opportunities for lobbying and participation in elections expanded substantially, thus suggesting a probable change in movement tactics after that time. Continuing shifts in opportunities and access to mobilizational resources during the post-Soviet period are also expected to affect movement development, as will be discussed in more detail below.

Identities and the Communist Experience

While the structural perspective has yielded significant insights into the organization, tactics, and developmental trajectories of late-communist social movements, little attention has been paid thus far to the question of mobilizational platforms. From a structural perspective, it has been argued that groups with platforms challenging party status and party goals are likely to be disadvantaged with respect to resource availability during the early reform period. In order to understand mobilization in late- and postcommunist societies, however, it is necessary to consider what kinds of platforms might be expected to have the greatest appeal to Soviet and post-Soviet constituencies.

I would suggest that the legacy of seventy years of communist rule would likely affect the mobilizational potential of competing platforms in two distinct ways. First, returning to my earlier characterization of the population of the Soviet Union and its successor states as resource poor, one might expect defensive platforms to have greater mobilizational potential than offensive platforms. This hypothesis arises out of studies of mobilization by resource-poor constituencies. Second, building on the insights suggested by the identity-oriented paradigm, platforms that can successfully appeal to strongly held group identities and identities reinforced by the communist experience are likely to have greater appeal to mass constituencies.

The linkage between level of wealth and preference for defensive versus offensive mobilizational platforms has been suggested by Tilly in his analyses of Western European social movements.[20] Mobilization is considered defensive when a group's collective action is in direct response to a threat to some valued aspect of its way of life, usually the material security and well-being of its members. In contrast, offensive mobilizations are based on a perception of new opportunities to realize common interests which lead group members to pool their resources and engage in collective action to improve their well-being. According to Tilly's study, the key determinant of whether a group engages in offensive or defensive collective action is its level of wealth and power.

While the rich and powerful are likely to concentrate their mobilizational efforts on offensive actions, the poor and powerless are much more likely to restrict their mobilizational efforts to defensive actions. The reasons for this dichotomy are twofold. First, because of their higher status in society, the rich and powerful are likely to be well protected from encroachments on their well-being and thus rarely find themselves faced with serious threats. In contrast, groups lacking financial resources and political clout are generally poorly sheltered from a myriad of threats to their way of life and material security. Thus, this group is much more likely to find itself in a situation where defensive action is required. Secondly, the decision to engage in collective action involves an estimate of the projected costs and benefits that might result from participation which is likely to deter poor, powerless constituencies from even considering offensive mobilizations. While the rich and powerful may, under some conditions, consider offensive mobilization to be likely to yield benefits far in excess of cost, this is rarely the case for less-well-off constituencies. Without power and resources, poor groups have little means to achieve offensive objectives, and engaging in such mobilizations is likely to put the little material security and status the group possesses in jeopardy.[21] Thus, in the absence of an immediate and tangible threat to some valued aspect of their way of life or material well-being, poor constituencies are unlikely to consider mobilization worth the potential cost.[22]

Outside the Communist nomenklatura, the remainder of Soviet society may be considered to have been both resource poor and disadvantaged in their power relationship to the state. Prior to the introduction of perestroika, groups outside the nomenklatura were barred from possessing significant stocks of land and capital, had little access to mobilizational resources, and were denied real channels for influencing political decision making. During the 1985–91 period, the Communist Party's privileged access to material and other mobilizational resources and monopoly control over political decision making slowly unraveled, thus diminishing the huge imbalances in wealth and power which characterized communist systems. Despite these changes, however, access by non-Communist actors to material and mobilizational resources remained extremely limited, and channels for influencing decision making remained inadequate.[23] It is thus reasonable to expect that during the 1985–91 period, much of Soviet society would favor defensive over offensive types of mobilization.

While defensive platforms were likely to have greater mobilizational potential than offensive in the late 1980s, those movements that also attempted to build on existing group identities would be expected to be especially potent. Such movements can not only reaffirm existing group identities but also provide a forum for people to explore and solidify these identities. What group identities might be available to provide the basis for mobilization in late- and postcommunist societies? Traditionally, class, profession, ethnicity, and territory are all considered likely bases for group identification. What must be asked in this case, however, is which of these social cleavages survived or were reinforced by the communist experience. It should be noted that while in retrospect it is now clear that national identity represented the most effective base for mobilization in the late-Soviet period, this was not at all obvious during much of the perestroika period. Attempts to mobilize mass movements on the basis of class, professional, or territorial identities were observed throughout the perestroika period, and it is thus important to take a serious look at the mobilizational potential of these group identities in late- and postcommunist societies.

Group identification based on class (defined here as common material interests and relationship to ownership of the means of production) was largely obstructed in communist societies. Because the state was the sole employer and monopoly owner of all land and capital equipment, all social groups outside the communist elite were similarly placed in relation to the means of production. It is thus extremely difficult to differentiate the material interests of one group relative to another. All groups were subjected to the domination of the state, and thus the cleavages that were reinforced by the structure of ownership are state versus society rather than workers versus bourgeoisie. While cleavages might well exist between workers, peasants, and intellectuals, they were not reinforced by the economic dominance of one group over another.

While communist institutions do not reinforce class identification based on common material interests, it is possible that identification along class or professional lines may have been strengthened by ideal factors. While there is some evidence that the regime occasionally fostered an antagonistic relationship between intellectuals and workers or peasants, there is little to suggest that any of these social groups developed strong group identities during the communist period. This is to be expected, given that (1) identification along professional lines was unlikely to have been carried over from the prerevolutionary period in which the bulk of society worked the land; (2) communist institutions did little to reinforce professional identities; and (3) Western experiences indicate that identification along professional lines is not generally a dominant social cleavage in modern societies.

While class and professional identities were neither inherited from the prerevolutionary period nor reinforced by communist institutions, many ethnic and national cleavages may be traced to the pre-1917 period. Furthermore, in contrast to class and professional identities, it may be argued that ethnic and national cleavages were often inadvertently reinforced by Soviet policies after 1917.[24] In order to simplify administrative tasks, republic boundaries established during the 1920s often followed ethnic and national lines, thus reinforcing a sense of group identity and distinctiveness. In addition, rather than viewing individuals as citizens of the USSR, passports identified people by "nationality," thus preventing anyone from losing sight of his or her primary national identification. Also, while all citizens were required to learn Russian as the official state language, republic languages were permitted to persist. Policies favoring the titular national group in education and in career advancement in the republics also served to reinforce national and ethnic cleavages during the Soviet period, particularly during the Brezhnev years.

While the party's goal was to create a new sense of Soviet identity which would transcend preexisting national identities, the extent to which it was able to succeed in this objective is questionable. In addition to inconsistencies in policy, the relative brevity of Soviet rule, particularly in some republics and regions of the USSR, hampered attempts to eradicate preexisting national identities. Repressive policies including bans on the use of non-Russian indigenous languages in official communications, prohibition of displays of national pride, and severe impediments to the propagation of national culture and history were aimed at undermining the maintenance of collective memories and shared group identities. Across much of the USSR, where several generations had been raised under such policies it might be expected that collective memories may have degenerated substantially over the USSR's seventy-year history. In areas such as West Ukraine and the Baltic republics, however, where the suppression of national identity had been pursued for little more than four decades and much of the population still shared memories of life before Soviet annexation, these repressive policies were likely to fall short of their full objectives.

As restrictions on speech and on association were gradually eased over the 1985–91 period, it is reasonable to expect that attempts might be made to resuscitate long-dormant national identities. In regions where national identity was well-developed at the time of incorporation into the union and subjection to communist repression was brief, this sense of a distinct group identity might be expected to rebound quickly, as has been the case in the Baltic republics. In other regions, however, where identity was poorly developed or decades of repression had cut deeply into collective memory, national identity was likely to present a much weaker base for group mobilization. In such regions, the development of a sense of group identity is expected to proceed slowly and with difficulty: language must be revived, culture and history must be explored, and a collective memory must be recreated. Thus, the role of ethnic or national identity is likely to vary considerably across republics in its ability to mobilize groups.

Finally, group identities based on shared territory rather than a sense of common membership in a national community must also be considered.

Although the administrative boundaries of the former USSR may be considered to have created and reinforced this group identity, it is unlikely to have the powerful appeal of a shared national heritage. Although common material interests may exist for members of a territorial entity, these groups lack the sense of national unity and shared history of their nationally defined counterparts. Thus, while available for possible mobilization, territorial group identities are not expected to form nearly as potent a base for the social movements emerging in the late-Soviet period as national and ethnic identities.

In addition to the fact that ethnic and national identities are expected to be the most powerful group identities to survive the Soviet period, these group identities are likely to play an especially important role in the mobilization of environmental and anti-nuclear movements. Because environmental movements stress the protection of land, territory, or a group of people, there is a natural affinity between environmental and national goals. This is particularly true in the case of nuclear power where the shoddily constructed and poorly operated nuclear power stations which littered the republics of the former USSR might easily be portrayed as powerful threats to the survival of a people or nation. Thus, anti-nuclear activism might be expected to benefit substantially from the mobilizational potential of national identity.

In sum, movements incorporating a powerful appeal for the defense of constituent interests and focusing on well-defined group identities—particularly national identities—are expected to have the greatest mobilizational potential in the late-Soviet period. While either factor alone might have substantial potential, combined they might well lead to a virtual explosion in social activism.

Mobilizational Patterns in the Post-Soviet Era

As discussed above, the resource mobilization and identity-oriented paradigms appear to provide substantial insight into expected patterns of mobilization during the perestroika period in the USSR. To recapitulate, resource scarcity is expected to lead to weak organizational forms, disruptive movement tactics, and a preference for defensive over offensive platforms. The unraveling of party control over resources and resulting uneven and constantly changing distribution of resources among competing groups in society is likely to lead to privileged access for movements perceived as nonthreatening to party interests. Furthermore, the existence of privileged movements might open the opportunity for more radical activists to use these movements as surrogates for less-acceptable causes. Finally, the possible persistence and reinforcement of strong national identities and the absence of competing group identities would create the potential for substantial mass mobilization for movements appealing to feelings of national community.

With the collapse of the USSR, however, one might expect mobilizational patterns to change substantially. The threefold legacy of resource scarcity, uneven distribution of mobilizational resources, and group identities reinforced by the communist experience is nonetheless still expected to play a significant role in shaping mobilizational patterns in the post-Soviet era.

Following the collapse of the USSR, resource scarcity is likely to be a continuing feature of the mobilizational environment in the Soviet successor states. During the first four years of independence, all of the newly independent states have faced immense hurdles in establishing their own, separate economies and moving from planned, communist economic forms to market capitalism. All of these countries have experienced severe economic hardships during this transitional period, and as of 1995, most (with a few exceptions) continued to show shrinking gross domestic products. Widespread poverty and lack of mobilizational resources would continue to favor movement profiles similar to those of the perestroika period.

Following the failed putsch of August 1991, the Communist Party was banned across much of the territory of the USSR. The expected result was a dramatic shift in mobilizational resources from the party to other competing groups in society. In reality, however, this shift has not been as substantial as expected. Despite the banning of the party, the nomenklatura elite were largely able to retain their privileged position with respect to mobilizational resources. Because the nomenklatura occupied most of the high-level positions in economic management in the country, these elites continued to have preferential access to such tangible resources as funds, meeting space, and communications technology. Similarly, the temporarily illegal but still existent network of the party, and its wealth of mobilizational experience, continued to provide these elites with such intangible benefits as organizational skills and networks for bloc recruitment. Conversely, nonparty actors continued to find themselves disadvantaged with respect to access to key mobilizational resources.

In addition to the hidden networks of communist elites maintaining a privileged position relative to resources, the state stepped into many of the newly independent countries to take over the party's control of the public realm. Particularly in the Central Asian states, the government maintained control over the press and limited the rights of independent associations. Even in the more successful democracies of Russia, Ukraine, and Belarus, government attempts to limit freedom of the press have not been uncommon. Thus, the role of the state in disseminating mobilizational resources to new groups in society has yet to be completely eliminated in most of the successor states.[25]

While the state may continue to play a role in shaping access to resources in the Soviet successor states, the collapse of its absolute domination in this area suggests that the days of surrogacy may be past. While state infringements on free speech and free association may be commonplace in the Soviet successor states, the level of repression of public activities is generally not high enough to force activists to hide behind surrogate causes. The severe penalties of the communist period have largely disappeared, and with promising mobilizational examples abounding in the former USSR and Eastern Europe, it would seem unlikely that radical activists would perceive the need to mask their causes behind surrogates.

While former communist elites and state actors have apparently maintained substantial advantages in their control of resources, two groups of nonstate actors should also be considered to be gaining privilege in respect to such access and control. The first are the new economic entrepreneurs and elites. A new class of wealthy business elites is emerging, and this group certainly has access to substantial resources of all kinds. Their potential for effective collective action is thus substantial. Their interest in environmental issues to date, however, has been minimal.

Those individuals and groups who can successfully link up with wealthy foreign individuals and organizations constitute the second privileged group in the post-Soviet era. While the ability of foreign individuals and organizations to establish bases within the former USSR or to channel resources to domestic activists was practically nonexistent prior to 1985 and quite limited during the perestroika period, the breakup of the USSR brought with it vast new opportunities for such activities. Thus, the most privileged groups of the post-Soviet era are likely to be either outgrowths of foreign organizations and movements or indigenous groups with strong financial ties to outside funders.

Finally, a dramatic change in the mobilizational potential of national identity is likely to be a critical factor in shaping mobilizational patterns in the post-Soviet era. 1991 marked the end of the nationalist struggle for fifteen of the dominant national groups of the former USSR. Political independence was achieved, and these national communities found themselves ensconced in their own nation-states. Thus, suddenly the powerful mobilizational potential of nationalism was taken away. These fifteen national groups now found themselves masters of their own fate. No longer could they rail against Moscow's imperialist domination. Their own national governments were now responsible for meeting the wishes of the people. The victory of nationalism meant that common grievances could no longer be combined with a nationalist critique of Moscow's continuing grip on these societies. A protest against local environmental contamination could no longer be construed as the complaint of a colony against its imperial master. It was no longer a cause which might mobilize an entire national community. Thus the national fragmentation of the USSR would be expected to substantially diminish the role of national identity in the mobilization of environmental and anti-nuclear activism.

Of course, with over 100 national groupings in the former Soviet Union, the creation of fifteen nation-states has not entirely buried the nationalist issue. In states with substantial national minorities, appeals to a national community might still have significant potency. One might thus expect that in national enclaves, such as Tatarstan (Russia) and Crimea (Ukraine), nationalism may yet play an important mobilizational role.

In sum, a continuing dearth of mobilizational resources for most independent actors is likely to discourage mass mobilizations. The absence of a powerful group identity to provide the basis for mobilization in the wake of the dissolution of the USSR will even further reinforce this tendency toward social apathy. Possible exceptions to this overall tendency, however, include new economic elites and groups sponsored by foreign actors. Furthermore, the potential for mass mobilization of national minorities in any of the fifteen newly independent states continues to persist.

Resources and Identity: *Understanding Patterns of Anti-nuclear Mobilization in the USSR and Its Successor States*

In reviewing the case of anti-nuclear mobilization during the 1985–95 period, it is clear that many of the expectations that emerged out of an analysis of resources and identity factors were in fact met. Interestingly enough, however, some divergences were also apparent. While detailed case studies of the anti-nuclear movements in Lithuania, Ukraine, Russia, and the national enclaves of Tatarstan and Crimea follow in chapters 2 through 6, a preview of these results will be useful in demonstrating the utility of the resource-identity framework as well as its shortcomings.

As predicted, the scarcity of mobilizational resources during the early communist period shaped the characteristics of all of the anti-nuclear power movements studied. Movements were loosely organized, with little linkage between territorially dispersed regions. In Russia, separate battles were fought around each station, with little or no communication between activists in different regions. In Ukraine, a Ukrainian umbrella organization was established, but communication and cooperation between regions remained low. As expected, in most regions a few dedicated activists directed all mobilizational activities, with supporters of the movement participating only in petition drives and mass actions. Outside this dedicated core, no stable membership existed. The tactics used were primarily disruptive in nature, with mass protests, strikes, and blockades dominating the movements' action repertoire. Interestingly enough, however, none of these anti-nuclear movements ever turned to violence as the ultimate disruptive tactic.

Mobilization on anti-nuclear platforms was also affected by the uneven distribution of mobilizational resources in society. While anti-nuclear protest was viewed as highly dangerous by Moscow decision makers, it appeared to be much more innocuous to local officials. Because local officials held no responsibility for energy production and distribution in the USSR, they saw little threat in the growing anti-nuclear movement. In fact, it appears that many viewed anti-nuclear activism as a safe outlet for popular frustrations. The anti-nuclear movement was largely viewed as a sign of growing environmental consciousness, and apolitical in orientation. Thus, local officials saw little reason to vigorously suppress anti-nuclear activism. In fact, as officials found themselves ever more dependent on popular goodwill for continuing their tenure in office, they often found it to their advantage to assist this highly popular movement. Thus, during the early period of perestroika, the anti-nuclear power movement was relatively privileged in its access to mobilizational resources. In contrast, at this time national and ethnic movements were viewed as highly threatening to the political system and thus were denied basic mobilizational resources and vigorously suppressed. Thus, ethnic and national groups found it much easier to mobilize on anti-nuclear platforms which incorporated their ideal interests, than to mobilize on purely national or ethnic platforms.

In all cases, the movements were initiated by a small group of intellectual elites. Writers and scientists often came together in their opposition to construction or operation of a nuclear power station in their territory. It is important to note that these intellectuals were rarely united in their motivations and objectives. Mobilization certainly carried a defensive material element, with numerous intellectuals displaying genuine fear of the possibility of another Chernobyl in their vicinity. Far more often, however, the discussions and publications of intellectuals referred to the ideal interests at stake. While some anti-nuclear proponents based their arguments on environmental principles, the overwhelming majority phrased their concerns in terms of nation, ethnicity, or territory. Nuclear power stations were viewed as a threat to the continued existence of a people or land, and these early activists couched their debate in terms of protecting and defending a particular group of people against the destructive policies of the center.

When censorship laws changed in January 1988 and these initiative groups could finally appeal to wider audiences, the calls to mobilize were entirely defensive in nature. Rather than calling on workers, peasants, and pensioners to fight for improved environmental protection or national sovereignty, anti-nuclear publicists portrayed nuclear power stations as horrifying and real threats to people's health and the continued existence of a national, ethnic, or territorially defined group of people. Plans for nuclear power stations were equated with policies of genocide, and calls for anti-nuclear mobilization contained a potent anti-imperialist element. Thus, these NIMBY (Not In My Back Yard!) movements, while mobilizing people on the basis of a threat to their material interests, were also able to take advantage of the overlap between anti-nuclear and national objectives to mobilize multiple constituencies on the basis of national platforms.

These calls to defend material and ideal interests of local populations were highly successful. Nonintellectuals responded in droves, some joining anti-nuclear associations, others signing and circulating petitions, and many participating in mass demonstrations or minor strikes. Because the appeal to this wider audience was based on ideal, as well as material, interests mobilization took on a distinctly national character in most regions outside the Russian republic.[26] Where there was a national identity that might be juxtaposed against Moscow (and implicitly, Russia), appeals to this national identity strongly reinforced the mobilizational appeal of the movement.

While all anti-nuclear groups called on local populations to protect themselves against a careless and uncaring imperial center, the linkage between the anti-nuclear issue and the community's sense of group identity varied. In some cases, such as Armenia and Lithuania, movement leaders clearly used the anti-nuclear issue as a surrogate for their true nationalist demands. As soon as these intellectuals perceived that mobilization on overt nationalist platforms was permissible, leading anti-nuclear activists in both of these republics quickly jumped into the new movements and soon thereafter abandoned the anti-nuclear cause. In other republics, such as Ukraine, anti-nuclear activists worked in association with the national movement but continued to lead a separate existence long after the nationalist movement had become well established. While the anti-nuclear movement took advantage of a Ukrainian sense of national identity, it would be inaccurate to call one a surrogate for the other. Finally, in Russian regions of the former USSR, the anti-nuclear movements became associated with calls for greater local decision-making power but were never linked to Russian nationalism. Thus, while our framework provides substantial insight into similarities between anti-nuclear movements in the USSR and its successor states, more attention needs to be paid to those factors that can differentiate movements. Why did the linkage between the anti-nuclear platform and national identity vary across republics?[27]

The impact of these proliferating anti-nuclear movements on energy decision making in the former USSR was both unexpected and substantial. Despite the organizational weaknesses of these new movements, the Soviet and republic governments were quick to bow to this unaccustomed public pressure. During the 1988–90 period, over fifty planned reactors were frozen or canceled, bringing the Soviet Union's nuclear power program to a virtual standstill. Decisions to cancel and suspend construction projects were taken on all levels—city, oblast, republic, and all-union—and ultimately lead to a serious energy crisis for the USSR and its successor states. The remarkable effectiveness of this grassroots movement opens interesting questions about the changing decision-making process during the perestroika period and after. While this question lies outside the scope of this study, I would suggest that the explanation for the movement's success lies not in its resources or organizational strength but rather in the Communist regime's weakening legitimacy. With the erosion of the Communist Party's integrity and sense of mission, the regime became unusually vulnerable to outside challenges.

While anti-nuclear movements achieved stunning successes during 1990, they unexpectedly stopped short of the ultimate goals of many activists. Rather than continuing the crusade and pushing for the closure of the dozens of operating reactors across Lithuania, Russia, and Ukraine, the movements withered away. By the end of 1991, none of the vibrant anti-nuclear movements of the perestroika period had survived. While some of the environmental organizations that were created to fight this battle still existed, mass mobilization on the issue was a thing of the past.

Following the breakup of the USSR in December 1991, an even more dramatic shift in popular sentiment became apparent. As the newly independent states struggled to establish themselves and build functioning independent economies, the impact of the decisions to curtail nuclear energy production became all too apparent. Energy crises in Lithuania, Armenia, Russia, and Ukraine quickly led to substantial changes in attitude toward nuclear power. Even the new political elites who had frequently emerged out of the local anti-nuclear and nationalist movements, began to express support for a reinitiation of local nuclear power programs. As moratoria and suspensions were canceled in Armenia, Russia, and Ukraine (and reconsidered in Lithuania), local populations remained apathetic on this issue. As expected, once nationalism and anti-nuclear demands could no longer be linked, the mobilizational potential of the anti-nuclear movement diminished significantly.

Since 1991, the indigenous anti-nuclear movements in all of the regions studied have dwindled to no more than a handful of still-concerned citizens (who mourn the loss of mass support for their agenda). Interestingly enough, as local interest in this subject has diminished, foreign concern about the safety of the former Soviet Union's nuclear power stations has stepped in to fill the void. The most active environmental organization in both Moscow and Kiev is now a regional branch of the international organization, Greenpeace. While staffed by members of the local population, these regional offices are funded almost entirely by the international organization. And rather than fitting the profile of a classical social-movement organization, these new groups bear closer resemblance to professional SMOs, with a paid professional staff and a sophisticated political strategy. Rather than relying on grassroots activism and protests, Greenpeace has favored political lobbying and using the courts to further its agenda. The emergence of foreign-sponsored, professional SMOs to replace the impoverished and largely apathetic indigenous movements represents a dramatic change in mobilizational patterns. It will be interesting to watch how the role of foreign-sponsored organizations evolves and what its impact may be on the growth of indigenous environmental consciousness and activism in the future.

As of 1995, however, a further change in movement characteristics could be distinguished. The inflow of capital from Western governments and nongovernmental environmental organizations interested in keeping indigenous environmental groups afloat caused a weak but discernible resurgence in environmental activism in some of the newly independent states. Whether this influx of capital was merely a futile attempt at "artificial life-support" (as one Ukrainian journalist has called it) or the stimulus needed to revive activism and spread environmental consciousness in these struggling countries remains to be seen.

By focusing on resource availability, our framework appears to provide significant insight into social movement characteristics typical of the late- and postcommunist periods in the Soviet Union and its successor states. It highlights the relationship between resources and opportunities on the one hand and movement profiles and development on the other. Because all of the movements studied emerged within the common structural context of a dying communist regime, it is not surprising to find that they shared numerous features. What is surprising, however, is the differences discovered between movements in different republics and regions. In fact, the role of national identity in mobilizing anti-nuclear activism was not a constant but rather varied across regions. The extent to which the idea of surrogacy accurately captures the relationship between the anti-nuclear and nationalist causes apparently differs across the republics and regions. In the upcoming chapters, it will thus be important to investigate more thoroughly the linkage between national identity and the rise and fall of local anti-nuclear movements. We will return to this question in the final chapter.

CHAPTER 2

Lithuania:
The National Element

Introduction

The anti-nuclear power debate in Lithuania centered around the republic's sole nuclear power facility, the Ignalina Atomic Energy Station (AES), located a mere 80 miles from Lithuania's capital, Vilnius. Plans for the Ignalina station were first introduced in the early 1970s. Initially, the station was to consist of two 1,000 megawatt (MW) RBMK reactors.[1] Gradually, however, the plans became more ambitious. By 1982, the proposed Ignalina station was to be the largest and grandest of the Soviet nuclear power facilities, consisting of four reactors of 1,500 MW each. In 1983, the first 1,500 MW reactor went into operation, with the second following in mid-1987, and the third targeted to go into operation by 1990.[2]

From the late 1970s, however, a scientific contingent within the Lithuanian Academy of Sciences strongly opposed the scale on which the Ignalina station was to be built and lobbied to limit the facility to a maximum of 3,000 MW. In the early 1980s, this group succeeded in channeling its concerns to the USSR ministries and forcing the establishment of a special USSR Academy of Sciences commission to review the plans for construction of the third and fourth reactors at Ignalina. This commission eventually conceded that a fourth reactor would meet neither the environmental or safety standards required and thus should be canceled. The commission refused, however, to back down on plans to build a third reactor. Thus the Lithuanian scientists won only a partial victory in their quest to scale back the grandiose plans of the central ministries.

What the scientists failed to achieve through inside channels, however, the populace accomplished quite dramatically in the summer of 1988. As new opportunities emerged for independent activities and voicing opposition to official policies, a movement opposing the expansion and even continued operation of the Ignalina Atomic Energy Station (AES) rapidly took form. Opposition to the Ignalina station was one of the first and most potent causes to mobilize the masses in Lithuania, and this outburst of popular opposition had a dramatic affect on official policy. By the fall of 1988, it was clear that construction of the third reactor at Ignalina could not be continued, and the project was indefinitely suspended.

In this chapter, I will consider (1) the role of intellectual elites in curtailing the expansion of the Ignalina AES, both before and after the introduction of perestroika and (2) how changes in political structures and the distribution of mobilizational resources affected the ability of intellectuals and society to effectively combat Moscow's plans to expand the Ignalina station. How successful were Lithuanian scientists in promoting their views through established inside channels? What role did scientists and other members of the intelligentsia play in mobilizing and directing the activities of the mass movement that emerged in 1988? How did changes in the distribution of mobilizational resources affect the characteristics and impact of anti-nuclear protest in Lithuania? What was the relationship between the mass-based anti-nuclear movement and the intellectually oriented Lithuanian national sovereignty movement? Finally, how did the achievement of national sovereignty and ultimately, independence, affect the attitude of both intellectual elites and the mass of society toward nuclear power in Lithuania?

Moscow Rules: *Pre-1986*

Lithuanian Scientists Battle Moscow[3]

Prior to the introduction of perestroika, decisions on nuclear power were the sole responsibility of Moscow decision makers. A variety of USSR ministries took part in nuclear power planning, construction, and operation, but republic and local organs had little or no say in any aspect of nuclear power decision making. At most, the central committee of a republic's Communist Party might be involved in vying for selection of a site in their republic for a new nuclear facility, since construction of such a grandiose project necessarily meant an inflow of funds and infrastructure to the republic. Republic scientists, however, were rarely consulted and approval by republic environmental commissions and other regulatory agencies was considered perfunctory; a republic's party chief simply requested approval from the relevant agency and was unlikely to meet any opposition on this demand. The idea that local residents might have some say in whether they wanted a nuclear power station next door to them would have been considered ridiculous in the pre-Gorbachev years. Moscow ruled, and the republics obeyed.

In Lithuania, the situation was no different. The decision to build the Ignalina nuclear facility came in the mid-1970s as part of Brezhnev's plan to dramatically expand the nuclear power program and construct new nuclear facilities across much of European USSR. The Ignalina site was selected by the USSR Council of Ministers after reviewing a number of possible sites in the Baltic republics and Belarussia. According to numerous reports of republic scientists, Ignalina was eventually chosen because of a railway line in the vicinity and a large lake which could be used for cooling—both factors expected to lead to reduced costs. The fact that Lake Druksai and the Ignalina region was a popular vacation spot for Lithuanians appears to have been completely irrelevant to Moscow decision makers.

The selection of the Ignalina site was apparently carried out with very little preliminary scientific research. In particular, detailed geological studies of the region were never undertaken[4]—a factor which would later come back to haunt Moscow. Republic scientists were given little role in selecting the site and designing the facility, even though a number of scientists from the Lithuanian Academy of Sciences opposed key aspects of the project from the very beginning. In particular, these Lithuanian scientists questioned (1) Moscow's decision to build RBMK-type reactors at Ignalina and (2) the overwhelming reliance on Lake Druksai to cool such an immense nuclear facility.

In 1976, the Lithuanian Academy of Sciences put together its first commission to consider the plans for Ignalina AES. The commission concluded that water-cooled graphite reactors of the RBMK design were in fact too dangerous to be constructed in highly populated areas and thus should be prohibited in the European USSR and Lithuania in particular.

This conclusion was adopted as the official position of the Lithuanian Academy of Sciences, and remained the Academy's position thereafter.[5] The commission's conclusions, however, had no impact on Moscow's plans to build dozens of RBMKs in European USSR and at Ignalina.

While the Lithuanian Academy of Sciences opposed Moscow on the selection of reactor design for Ignalina, the key point of contention between Lithuanian scientists and their counterparts in Moscow lay in the question of Lake Druksai's capacity to adequately cool the Ignalina station. Initially, the Ignalina station was rather modestly designed to produce a mere 2,000 MW of power; in the late 1970s, this was increased to 3,000 MW. Lithuanian scientists, however, were immediately concerned by the capacity of Lake Druksai to cool the facility without sustaining environmental damage to itself. In the late 1970s, members of the Lithuanian Academy of Sciences concluded that the lake could only sustain a maximum of 2,500–3,000 MW of power production, thus making Moscow's plans for two 1,500 MW reactors questionable. In a daring 1980 article (and the only one of its kind to be published in the Lithuanian press prior to perestroika), three Lithuanian scientists criticized plans for using Lake Druksai to cool Ignalina, suggesting that this cooling design was inadequate and should be reconsidered.[6]

The concerns of the Lithuanian scientists, however, apparently had little impact on the Moscow *atomshchiki* (the common term for members of the official nuclear establishment). In 1982, without prior warning or discussion with members of the Lithuanian Academy of Sciences, Moscow announced its decision to expand the plans for Ignalina to 6,000 MW. The Lithuanian Academy of Sciences was quick to react, forming an official commission to review Moscow's expansion plan and to offer its own recommendations. The commission, headed by vice president of the Lithuanian Academy of Sciences, Algirdas Žukauskas, concluded that the expansion was not permissible and sent its recommendations to the Lithuanian government.

The Lithuanian government, however, returned the recommendations to the commission, suggesting that the Lithuanian scientists would be more successful if they focused on convincing their scientific counterparts in Moscow. Thus, it was openly acknowledged that the Lithuanian government played no significant role in nuclear power decisions in their own republic. Such decisions were the domain of Moscow alone. The scientific commission put together a report and forwarded it to the president of the USSR Academy of Sciences, one of the USSR's leading proponents of nuclear power, A. Alexandrov.

By working through inside channels and focusing on convincing Moscow's scientists of the validity of their concerns, the Lithuanian scientists succeeded in putting the Ignalina-expansion question on the official agenda of the relevant Moscow ministries and several concerned institutes. Numerous Moscow scientists from a variety of institutes concerned with nuclear power engineering have noted heated debates held on this issue in 1983 and 1984. A number of special interdepartmental conferences were held on the question of Lake Druksai and its capacity to cool the Ignalina station without sustaining environmental damage.[7]

The concerns of the Lithuanian scientists appear to have been legitimate and their arguments, convincing. As a result of the USSR Academy of Sciences review of the expansion project, the Academy's president, A. Alexandrov, agreed to limit Ignalina to a mere three reactors (i.e., 4,500 MW). The Lithuanian scientists, however, felt that their victory was incomplete. Their goal had been the cancellation of the third and fourth reactors, and their achievement fell short of this. The Lithuanian Academy of Sciences remained officially opposed to the construction of a third reactor at Ignalina from that time forward.

Public Activism

While scientists at the Lithuanian Academy of Sciences were mobilizing to fight the construction of the world's largest nuclear facility, the public remained both docile and ill-informed on the issue. Prior to 1988, critical discussion of nuclear power plans was strictly forbidden in both the central and republic media. While articles on Ignalina were published in Lithuania, they always described the station as "the project of the century," the ultimate nuclear power station and an object of well-deserved pride for all residents of Lithuania. Articles were published praising the construction teams for beating deadlines and raising the reactors so quickly. With one minor exception, no critical articles on the Ignalina AES can be found in either the Lithuanian or central press.[8]

The result of this lack of glasnost was a society remarkably sheltered from the anti-nuclear sentiment that had infected much of the West in the 1960s and 1970s. Nuclear power had always been praised as absolutely safe and clean,[9] and there is no sign that any significant sector of society (outside the scientific establishment) ever even questioned the party line on nuclear power. Interviews with environmental activists of the late 1980s confirm that neither they nor society at large felt any significant concern about the possibility of a nuclear accident in their republic. The government, not the people, was assumed to be responsible for ensuring nuclear safety, and most people simply digested the propaganda handed to them and trusted that the government's promises were sound.

In addition to little indication of anti-nuclear inclinations prior to perestroika, the opportunities for society to mobilize in opposition to official policy were strictly limited. Effective mobilization requires, at minimum, that people be provided with opportunities to freely associate outside the control of the state and to openly exchange their ideas. Without freedom of association and speech, popular mobilization can be at best sporadic and disorganized. But, as noted earlier, in communist societies the party monopolizes the public realm and prevents oppositional forces from effectively communicating with each other and publicizing their concerns. Prior to the introduction of perestroika, the party controlled all branches of the media and, through both official and unofficial (i.e., self-) censorship, were able to effectively dictate which opinions were acceptable for public consumption.

The nuclear power issue was considered to be a particularly sensitive one, closely linked to the military security of the Soviet Union. Thus, articles and broadcasts that referred to the Soviet Union's nuclear power program required permission not only from the routine censorship organs (i.e., Glavlit) but also from the military censor in Moscow. Clearance for the publication of an article on nuclear power thus took at least a month to receive and was rarely given for any article with even a hint of skepticism about the wisdom of the USSR's nuclear power policy.[10] An indication of the lack of glasnost exhibited in the nuclear power issue can be found in the fact that prior to 1988, the decision to cancel the fourth reactor at Ignalina (which was apparently reached in 1984) was never reported in the Soviet press. As late as 1987, neither the Soviet public nor Western specialists studying the Soviet Union's nuclear power program were aware of the decision to cancel reactor #4.[11]

In addition to being denied the opportunity to communicate concerns about nuclear power and force this issue onto the public agenda, Soviet citizens were also denied the possibility of forming independent clubs and associations to mobilize on this issue. All forms of association outside the household were monopolized by the party-state, and independent environmental or anti-nuclear clubs were strictly prohibited.

The state sponsored its own version of environmental associations for people wishing to participate in this issue area. The foci of such associations and clubs, however, were always completely unthreatening to official state policy. Thus, *druzhinas* (student environmental clubs) at universities generally focused on environmental education, campaigns against poaching, illegal Christmas-tree thefts in state forests and parks, and such politically innocuous themes. Such official organizations were highly restricted in their activities and rarely directed their attention to big industry and the state's role in the destruction of the natural environment. There is no sign that the nuclear power issue ever made it onto the agenda of any such official organization prior to 1988.[12]

Without protected rights of association and free speech, the only route open to a determined activist was illegal dissent. While a small dissident movement did exist in Lithuania prior to the introduction of perestroika, this group of people was limited in number and able to reach a very tiny audience at best. Moreover, the dissident movement in Lithuania was only marginally concerned with environmental issues. The main focus of Lithuanian dissent was on religious freedom and Lithuanian national survival. While an occasional reference might be made to Moscow's destruction of the Lithuanian environment, the dissident movement in Lithuania was certainly not established around environmental questions, and there is no indication that nuclear power ever played a role in dissident debates.[13]

Thus, outside the halls of the Lithuanian Academy of Sciences and other elite scientific establishments in the republic, the remainder of society was largely ignorant and unconcerned about the colossal nuclear power station being built in their republic. The party-state's monopoly control over all sources of information and forms of public association virtually guaranteed that few people would question the wisdom of Moscow's nuclear power policy, and that the few who might question would be deterred from voicing such concerns in any kind of public forum. Public mobilization on this issue was thus a virtual impossibility prior to the radical restructuring of the Soviet system that began with Mikhail Gorbachev.

The Effects of Gorbachev, Chernobyl, and Perestroika

The Early Years

The dramatic mobilization on the nuclear power issue which took place in the late 1980s across the entire USSR can be traced to two primary factors: Chernobyl and perestroika. The former provided an obvious grievance and trigger for the movement. The latter made the movement possible by providing the opportunities and resources for effective mobilization outside the auspices of the state. Thus, in the late 1980s both motive and opportunity came together to presage the emergence of anti-nuclear activism.

In Lithuania, as elsewhere, glasnost was rather slow to come to the nuclear power issue. Military censorship of all nuclear power articles and broadcasts remained in place until January 1988, and discussion of this issue was strictly limited during the early period of perestroika.[14] Prior to 1988, the Ignalina AES, when mentioned, was always referred to in highly positive terms, and few details were divulged about the work at the station. After the Chernobyl accident, references to construction at Ignalina in the Lithuanian press became even less common, with no mention of the status of reactor #2 after April 1986. When reactor #2 finally went into operation in August 1987, it was accompanied by none of the usual fanfare and no publicity.[15] As noted earlier, even the decision to cancel the fourth reactor was not publicized until April 1988. Thus, despite the much-touted policy of glasnost, most residents of Lithuania remained largely in the dark about the nature and potential threat of the immense power station being built in their republic.

Another indication of the lack of openness regarding the nuclear issue may be found in the printed discussion of the Chernobyl accident, when it appeared, in the Lithuanian press. Interestingly enough, Lithuanian publications did not attempt to publish their own analyses of the situation but relied solely on the central press as their source. Articles on the accident were drawn almost exclusively from *Pravda* and *Izvestiya*. During the months immediately following the accident, only one article was published in the Lithuanian press addressing the question of potential radiation exposure in the Lithuanian republic. This was an interview with the Lithuanian minister of health, Jonas Platukis, who assured the public that radiation levels were being monitored in Lithuania, but there was nothing to worry about. By the summer of 1986, coverage of Chernobyl in the Lithuanian press trailed off to almost nothing.[16]

The party seemed to believe that by holding the lid on public discussion, it could prevent anti-nuclear opposition from coalescing. And in fact, this supposition appears to have been well founded. Although the Chernobyl accident may well have generated some concern about the safety of Ignalina AES among Lithuanian residents, such concerns were not permitted to overflow into the public arena. There have been rumors that the Lithuanian Central Committee received a number of letters in the spring of 1986 expressing concern about Ignalina. Letter writing, however, had always been an acceptable outlet for minor grievances and was nothing new to the communist system. Thus, while local citizens began to develop an awareness of a potential threat in their region as a result of Chernobyl, the lack of opportunity to publicize and discuss such concerns precluded mobilization on the issue.

Prior to the spring of 1988, virtually no strong criticism of the Soviet Union's nuclear power program in general or the Ignalina AES in particular were published in the high-circulation Lithuanian press. In addition, the informal (i.e., independent, non-state) press did not begin to appear in any significant quantities until 1988; thus the potential opponents of Ignalina and nuclear power had no media forum in which to publicize their concerns. While scientific opposition clearly existed, the scientists remained isolated from the public, and few possibilities existed for them to present their views to society and to attempt to mobilize people against nuclear power.

While the scientists were unable to appeal to a mass audience prior to 1988, this did not mean that they remained dormant. In fact, scientists were beginning to organize themselves into a network of environmentally concerned citizens. While opposition to the Ignalina expansion project in the early 1980s undoubtedly created linkages between diverse scientific communities in Lithuania, the process of building an environmental network did not end there. In 1985, the Soviet government unveiled a plan to begin oil drilling along the Baltic shoreline, close to a popular vacation spot near Neda. The scientists were quick to respond, pointing out the environmental hazards associated with this project.

In the new spirit of glasnost, however, this issue did not remain the sole domain of the scientific community. Rather than working exclusively through inside channels, as had been the case with the opposition to Ignalina's expansion, the scientists began to form a coalition with other members of the intelligentsia, particularly writers and artists. While opposition to the project remained an elite movement, which did not incorporate sectors of society outside the intellectual realm, it provided an opportunity for these intellectual activists to forge networks and gain useful mobilizational experience.

The success of this elite movement in achieving its objectives may also have played an important role in the later mobilization of the green movement in Lithuania. In 1986, the USSR government responded to this pressure from the elite by canceling the Baltic drilling project. Thus, these intellectuals learned that their opposition could have a real effect on Moscow's policies and were rejuvenated with new confidence and by a sense of efficacy in their efforts. Most environmentalists that I interviewed in 1991 pointed to this opposition movement as the root of the green movement that later emerged.

In opposing the Baltic drilling project, however, intellectuals went beyond mere scientific concerns. While the environmental hazard was the key focus, opposition to the project represented an opportunity to voice resentment over Moscow's colonial treatment of Lithuania. By ignoring the fact that Neda was a popular resort area, Moscow had displayed an overt insensitivity to republic priorities, and it was this, rather than the question of drilling in itself, that probably led to the formation of a coalition between the scientific and cultural intelligentsia.

Žemyna, the Goddess of the Earth

While changes in the opportunities to appeal to mass audiences and to form associations autonomous to the state were slow to materialize during the first few years of perestroika, this changed dramatically in Lithuania during the spring and summer of 1988. When 1988 began, access to the official press was still strictly limited; the unofficial press was as yet unborn and independent associations and clubs were almost nonexistent. Despite Gorbachev's calls for glasnost and the radical restructuring of state-society relations, very little had changed in Lithuania as of December 1987.

1988, however, was a year of testing boundaries and pushing the limits of independent activity ever further. And as people began to learn of new opportunities to mobilize independently and even in opposition to the state, the entire process accelerated rapidly. What began in February 1988 as rather tentative feelers on the possibilities of free speech and independent organization, ended with mass mobilizations and calls for the all-out secession of Lithuania from the Union.

In late 1987 and early 1988, members of the intelligentsia began meeting privately and considering the possibility of forming independent associations focusing on issues of concern to them. Among those early activists were scientists concerned with environmental problems in Lithuania. Approximately ten young scientists from the Lithuanian Academy of Sciences and the University of Vilnius, most of whom had worked together in the earlier battle against the Baltic drilling project, began holding informal meetings in late 1987 to discuss what they considered to be the most serious environmental threats in the republic. At the top of their agenda was the hazards and environmental impact of the ever-growing Ignalina AES. These young physical scientists were to form the core of one of the first independent associations to emerge in Lithuania: the Žemyna club.[17]

The Žemyna club was founded in late 1987 and described itself as an independent environmental association. From the start, its members agreed that the first order of business would be to address the safety and environmental concerns associated with the Ignalina station. At the time the club was founded, however, the creation of an association completely autonomous of the state and opposing the state on such a critical environmental issue, was considered to be an extremely bold and daring feat. In fact, when the club attempted to register its existence under the auspices of an officially recognized body in the fall of 1987, it was initially rejected by all of the organizations it appealed to due to fears of a strongly negative government reaction to the group.

In December 1987, the Žemyna club finally succeeded in registering itself under the auspices of two comparatively progressive and sympathetic official organizations: the Lithuanian Komsomol and the Presidium of the Lithuanian Academy of Sciences. The scientists were in fact quite lucky to find support for their cause at the top levels of the Lithuanian scientific community. While most of the members were quite young, in their late twenties or early thirties, the club included several figures of significant scientific stature: an institute director and a vice president of the Lithuanian Academy of Sciences.[18] Moreover, while membership in the club was concentrated among junior scientists, sympathy for its cause was widespread throughout the scientific community and particularly within the Presidium of the Lithuanian Academy of Sciences.[19]

Thus, the Presidium of the Lithuanian Academy of Sciences was able to provide a certain degree of support and shelter for the group which was significant in allowing the association to grow and spread its message to society. The founding congress of Žemyna was held at the Academy on February 11, 1988, and the club was provided with its own room at the Presidium of the Academy. Furthermore, during the spring of 1988, it was permitted to hold weekly discussion meetings and open forums at the Academy.

During the spring of 1988, the Žemyna club grew to about thirty-five members and was able to spread its concerns about Ignalina to a substantial portion of the population. In February 1988, Žemyna members began a press campaign against the Ignalina station. This was led by Žemyna's most active and committed member, a young physicist named Zigmas Vaišvila. Vaišvila wrote a series of articles on the dangers of Ignalina and succeeded in having them published in the official newspaper of the Komsomol.[20] These articles were considered daring and ground breaking in their candor and ignited widespread discussion of the Ignalina issue in Lithuania.

In addition to the press campaign carried out by Vaišvila, the Žemyna club hosted a number of open discussion meetings to focus the public's attention on environmental issues in Lithuania. In March and April, two particularly large public meetings were conducted by Žemyna on the question of the Ignalina station. These sessions were attended by approximately 400 people and included speeches by leading members of the Lithuanian scientific community.

While the ostensible reason for the creation of the Žemyna club and the hosting of such public forums was to address environmental issues in

Lithuania, the real story went far beyond environmental concerns. People who attended the founding congress of Žemyna in February 1988 recall that the congress was not simply an occasion for addressing environmental problems but also an unprecedented opportunity for the founders of Žemyna to vent their frustration with Moscow's treatment of Lithuania. The congress was seen by many as a political outlet and a real turning point in the activation of the Lithuanian intelligentsia. Rather than focusing on mere health and safety concerns, Žemyna members decried Moscow's dictatorial treatment of Lithuania and their own impotency in caring for the Lithuanian environment and people.

This political activation was spread even further when Žemyna opened its doors to the public in March and April of 1988 and hosted the open forums on Ignalina. Again, while the discussion focused on the nuclear power station, the participants in the forums used the opportunity to lambast Moscow for its destruction of the Lithuanian environment. These forums were unprecedented in their size and in the boldness of discussion and represented a critical turning point in the politicization of Lithuanian society.

The initial reaction of the Lithuanian government and the party to the formation and activities of the Žemyna club was somewhat muted but definitely unenthusiastic. Žemyna members report that the party made a number of attempts to thwart their activities during the winter of 1987–88. The party actively attempted to prevent the group's registration during the fall of 1987, and to shut the club down. The official sponsorship by the Presidium of the Lithuanian Academy of Sciences, and the Komsomol, however, prevented the party from taking any overtly aggressive steps against the club.

As Žemyna's appeal rapidly spread throughout society, and green clubs began sprouting throughout Lithuania, however, the party took steps to defuse the Ignalina situation. Thus, a roundtable of specialists, environmentalists, and other intellectual activists was organized by the Lithuanian party and held at the Ignalina station in March 1988. The purpose of this roundtable was to reassure the public of the safety of Ignalina and thus defuse the growing opposition movement. The actual achievements of this session, however, fell far short of its organizers' objectives.

The roundtable was led by the chief engineer of the Ignalina station,

Gennady Negrivoda, who fielded questions about the plant and explained how its safety systems worked and what improvements had been made on the station since the Chernobyl accident. Negrivoda, however, found that his audience was less than willing to listen and accept his reassurances. The session was attended by a number of outspoken environmental activists as well as established scientists who had long-opposed the Ignalina project. Vice president of the Lithuanian Academy of Sciences, Algirdas Žukauskas, who had earlier chaired the commission to evaluate the expansion of the Ignalina AES and led the Moscow lobbying effort against the project in the early 1980s, spoke out against Moscow's monopoly on information about Ignalina and the failure to include Lithuania's scientists in decision making on the station. Romualdas Lekevičius, director of the eco-genetics laboratory at University of Vilnius, spoke even more heatedly against the station and claimed that Ignalina was spewing radiation into Lake Druksai and that the radiation was already producing mutations in the lake. By all reports, the audience was far from the docile group expected by Ignalina's administration. Nonetheless, intellectuals attending the roundtable complained afterwards that the station administrators had dominated the session and had not been willing to listen to their concerns.

In an effort to appease the new opposition, the deputy minister of the Lithuanian Ministry of Land Reclamation and Water Resources, Julius Sabaliauskas, announced to the group that Moscow had recognized the inability of Lake Druksai to provide adequate cooling for four reactors and had thus canceled the fourth reactor. Though this decision had come years before, this was the first the public had heard about it.

Further attempts were made to allay public fears by publishing two reports on the roundtable in the main party newspaper, *Tiesa*.[21] These reports focused on the safety improvements at Ignalina and officially announced the cancellation of reactor #4. The articles were not, however, a stunning example of glasnost, since the concerns of the anti-Ignalina contingent were barely mentioned. In fact, throughout the spring and summer of 1988, *Tiesa* remained a conservative stronghold of the party and largely closed to progressive or oppositional viewpoints.

In March, Žemyna also undertook to demonstrate the level of public support for its cause by beginning a petition campaign against Ignalina. The campaign was carried out very informally, with both members and nonmembers circulating petitions of their own composition. Thus, some petitions called for closing Ignalina, others for halting the expansion project, and others for an international inspection of Ignalina's safety standards. By late spring, however, Žemyna had collected approximately 44,000 signatures opposing the Ignalina station, which they then sent by car to Prime Minister Nikolai Ryzhkov in Moscow. Ryzhkov's response, however, was simply another reassurance that everything was okay and Moscow had everything under control.

In April, Žemyna members decided to expand their petition drive by soliciting signatures through the republic press. They published an article asking people to send their opinions about the Ignalina station to the offices of *Komjaunimo tiesa*. According to Žemyna members, over 70,000 signatures opposing the Ignalina station were received during April and May. The activists called this an "informal referendum" and used this display of popular support in their attempts to sway the authorities in Vilnius and Moscow.

The informal referendum seems to have had a dramatic impact on the Lithuanian government's platform regarding Ignalina. In February and March, the Lithuanian authorities were struggling to defuse the anti-Ignalina movement; by late spring, the implications of the popularity of this issue began to sink in. Suddenly Lithuanian authorities began to recognize the political value of shifting to the other side. The anti-Ignalina movement brought home to the Lithuanian authorities the level of popular dissatisfaction with Moscow and the Communist Party, and provided them with a tool for shoring up their own self-image relative to both Moscow and the CPSU (Communist Party of the Soviet Union). Thus, by May 1988, the Lithuanian authorities had clearly begun to shift their stand on Ignalina to the side of the environmental activists.

National mobilization[22]

While the growing opposition to the Ignalina AES which emerged during the spring of 1988 certainly reflected popular concerns about nuclear safety and the environment, these were not the only factors involved in mobilizing people against the station. In appealing to people to fight Moscow's nuclear energy policy in Lithuania, environmental activists appealed to more than simply environmental sympathies; they called on people to save the Lithuanian nation and the Lithuanian people from potential annihilation. Throughout the spring and summer of 1988, the Ignalina nuclear power station was consistently portrayed by members of Žemyna and other activists as a threat to the survival of Lithuania. They warned that an accident at Ignalina would contaminate the entirety of the tiny republic and force its residents to relocate across the USSR. An accident would mean the end of Lithuania, the dispersal of Lithuanians as a people, and the death of their dreams of nationhood.

In portraying Moscow's nuclear power policies in Lithuania as a potential act of "genocide" and warning the public of the threat of Lithuanian "extinction," the leaders of the Žemyna club were clearly appealing to the Lithuanian national identity. They explicitly called on the population to defend what they held dear: their nation. If one looks at the press campaign against Ignalina that was carried out during the spring and summer of 1988 or the reports on the open forums held by Žemyna in March and April, it is clear that the anti-nuclear activists were consciously attempting to bind the anti-nuclear and national causes together. Whether their goal was to use national identity as a tool to mobilize people against nuclear power or vice versa, however, remained to be seen.

In early June of 1988, the anti-nuclear movement underwent a dramatic change in orientation and organization. Up until that time, opposition to Ignalina AES had been led by members of environmental groups, particularly the Žemyna club, and had been largely based within the intelligentsia; in June, leadership of the movement shifted to a new independent group, and anti-nuclear opposition began to take on the characteristics of a mass movement. This change came with the formation of the Lithuanian Movement for Perestroika on June 3, 1988.

During the winter of 1987–88, scientists opposing the Ignalina nuclear power station were not the only group among the intelligentsia to begin to organize themselves. Other associations of intellectuals were also emerging and beginning to play a public role in society. Groups within the Writers' Union and other artistic unions were gathering for discussions on issues of concern to them. These issues were often linked to Lithuania's history, identity, and place in the USSR. New and more progressive associations were beginning to splinter from the old, official organizations. For example, a group of progressive economists broke away from the official Znaniye club to form their own, independent Economists' Club.

Other independent clubs of intellectuals sprouted throughout the spring of 1988.[23]

Thus, when the Lithuanian Communist Party chose to select delegates to the Nineteenth Extraordinary CPSU Conference in June 1988 according to the old procedures—that is, at a closed session of the Central Committee—the newly organized intellectuals did not accept the outcome docilely. They were, in fact, outraged that the list of delegates consisted of the usual bureaucrats and included only three members from the scientific and cultural intelligentsia. After all the propaganda about glasnost, perestroika, and democratization, this antidemocratic and secretive method of selecting conference delegates seemed to expose the hollowness of the entire reform process.

At the initiative of the Žemyna club, members of Žemyna, the Economists' Club, and several other informal groups met on May 30 at the Lithuanian Academy of Sciences to discuss their frustration with the democratization process in Lithuania. At that meeting they learned that the Presidium of the Lithuanian Academy of Sciences had established a commission to make recommendations for a new Lithuanian constitution, and decided that the informal groups should request a meeting with the chairman of that commission and attempt to influence its recommendations. At this session, Žemyna members and other intellectuals also recognized the need to organize themselves under a single umbrella in order to strengthen their position in shaping Lithuanian politics.

On June 3, members of Žemyna and other informal groups met with the chairman of the constitutional commission, Eduardas Vilkas, in the Academy of Sciences conference hall. The meeting was a lively one, with over 500 intellectuals—most of them frustrated with the progress of reform in Lithuania—in attendance. According to reports of the meeting, Vilkas rapidly lost control over the session, and what followed was a free-for-all of complaints, criticisms, and new ideas about reforming the system. It was Zigmas Vaišvila, the head of Žemyna, who then called on people to follow Estonia's lead and form a popular front to unify the independent democratizing forces. An initiative group to lay the groundwork for a new Lithuanian Movement for Perestroika was immediately elected. While some members of the group denied that the composition of the initiative group had been predetermined, others acknowledge that at a private meeting of informal group leaders on June 2, a preliminary decision was made to form the Movement and a tentative list of members was drawn up in preparation for this contingency.

Thus, the Lithuanian Movement for Perestroika, commonly known as Sajudis, was born, and the struggle against the Ignalina nuclear power station took a new turn. Though the Žemyna club continued to exist and participate in anti-nuclear activities, most of its most active members shifted their attention to Sajudis, and it was Sajudis that organized future mass actions against the Ignalina station. The chairman of Žemyna, Zigmas Vaišvila, immediately became the chairman of the Ecology Commission of Sajudis and, from that position, continued to be the most dynamic player in the anti-nuclear campaign.

The anti-nuclear movement not only did not suffer from the shift in leadership from Žemyna to Sajudis, it prospered. In fact, the campaign against Ignalina became one of the early centerpieces of the Sajudis movement. It was discovered early on that the Ignalina issue was a very powerful mobilizer in society and one of Sajudis's most potent tools for calling people onto the streets. Thus, Sajudis used this issue during the summer of 1988 to wake society from its lethargy and drag it into public activism. Anti-nuclear activism was seen as a first step in politicizing society and mobilizing people to participate in the Lithuanian national cause. The anti-nuclear movement became a building ground for the creation of mobilizational networks and the resuscitation of a sense of Lithuanian national identity.

In addition to being a useful tool for mobilizing a long-dormant society, the anti-nuclear cause had another mobilizational advantage. Because opposition to the Ignalina nuclear power station could be categorized as environmental, it was not viewed as an overtly political movement nor as a threat to the communist regime. Thus, mobilization on this issue was viewed as much safer than other, more political platforms. For the entrepreneurs organizing the movement, this meant that they could go ahead with their activities without fear of massive retaliation from the authorities. For the society being mobilized, this meant that they could participate in the movement and gain a sense of their own public role and political efficacy without the fear that they would be severely punished. Thus, the anti-Ignalina cause suited both its leaders and society.

For intellectuals and the mass of society, the anti-Ignalina movement had different meanings and purposes. For those intellectual entrepreneurs leading the movement, it was undoubtedly a surrogate for the demands of national sovereignty that were as yet too dangerous to be made. It was a chance to fight the colonial relationship with Moscow without overtly threatening the regime. For the mass of society, however, opposition to Ignalina represented a real fear of a nuclear disaster, like Chernobyl, combined with the desire to defend the Lithuanian nation. It was a desire to prevent further contamination in Lithuania by radioactive substances or by the incursion of the Russian presence.

It must be kept in mind that the anti-Ignalina movement, in fact, incorporated a strong anti-Russian dimension. The Ignalina nuclear power station had been built by Russians and was staffed primarily by Russians. A special town had been constructed near the station to house these migrant Russians. Thus, Ignalina and the town of Sniečkus represented not only the dangers of nuclear power but also the threat of "national contamination."[24]

Thus, the Ignalina issue had a multiple appeal that vastly enhanced its mobilizational potential. For some it was a surrogate for nationalist protest, for others an environmental threat that had to be combated. But for most of society it was simply a way to voice frustration with Moscow's colonial treatment of Lithuania and with the threat of Russia overwhelming what was left of their Lithuanian heritage. It was the first step on the road toward politicization and active participation in the movement, one that would eventually lead to full independence from Moscow and Russia.

Throughout the summer, anger with the poor safety record and continued expansion of Ignalina AES grew. Anti-nuclear and environmental clubs proliferated across Lithuania. Petitions continued to be circulated, and Sajudis sponsored several bike hikes around the republic to publicize the cause.[25] By the end of 1988, there were over fifty environmental clubs in Lithuania, and many have described their creation as the beginning of the "rebirth of the Lithuanian nation." Public debate also expanded as Vaišvila and other anti-nuclear activists gained more access to the press.

Recognizing the popularity of the anti-Ignalina platform and the need to shore up its image in society, the Lithuanian Communist Party began to take steps to demonstrate their sympathy for the cause in June 1988. Their first action was to officially request that construction of reactor #3 be suspended until an appropriate safety review had been carried out. The Lithuanian authorities complained that no final construction plan had ever been approved for the third reactor and that the appropriate water resources agencies and atomic control commissions had never been consulted. Their petition for a review was sent to Prime Minister Ryzhkov in June, and their demands were almost immediately rejected as unnecessary.

In July, the Lithuanian Communist Party again went on record as opposing the expansion of Ignalina. Communist Party first secretary Ringaudas Songaila, a man not known for his progressive views or support of the democratization movement, spoke out against the construction of reactor #3 at the Nineteenth Extraordinary CPSU Conference in Moscow. Although Songaila did little more than mention the request and did not pursue the demand in any forceful way, the fact that he brought it up at all is remarkable.

When the delegates returned to Vilnius, a public rally was held on July 9 to hear the results of the conference. It has been estimated that over 100,000 people attended this meeting. During his presentation to the crowd, Central Committee secretary Algirdas Brazauskas announced that the Lithuanian government had halted its financing of the third reactor at Ignalina and would withhold further financing until construction plans had been properly approved.[26] This announcement, of course, generated a very positive reaction from the crowd and helped solidify Brazauskas's reputation as one of the most progressive and sympathetic members of the Lithuanian Communist Party's leadership. At the end of this meeting, Vaišvila proposed a resolution against further construction at Ignalina; the crowd informally affirmed his proposal.[27]

In late July, the Lithuanian government responded to popular opposition to Ignalina by unilaterally terminating Lithuanian financing for the third reactor. The USSR Ministry of Atomic Energy, however, appealed to the USSR Council of Ministers to override this decision, and financing from Moscow continued.[28] According to Lithuanian reports, at least 50 percent of the construction forces were still at work at Ignalina in August 1988. As noted later, however, the Lithuanian government's decision to halt financing of the project was largely a symbolic one, since only 3,000 rubles per day were lost as a result of this decision.[29] It is, however, indicative of the growing assertiveness of the Lithuanian government relative to Moscow and the growing recognition by Lithuanian political figures of the importance of basing their support in their own population rather than relying on Moscow to sustain their authority.

Finally, on August 25, the Lithuanian government organized a major roundtable at the Ignalina station to resolve the question of the fate of reactor #3. While the March roundtable had been organized by the Lithuanian government in an attempt to defuse public fears about nuclear power, the August session differed substantially in that the Lithuanian government had now clearly swung over to the side of the Lithuanian population and was ready to engage in battle with Moscow to prove its loyalty to Lithuania. The roundtable was attended by representatives of the Lithuanian government, the relevant Moscow ministries, the IAES administration, Lithuanian specialists, and environmental activists.

The Lithuanian government and scientific establishment clearly opposed continuation of construction of reactor #3, while the director of the Ignalina station, Kromchenko, and representatives from Moscow strongly favored continuation. Thus the discussion was a heated one and lasted over five hours. Kromchenko argued that over 124 million rubles had already been invested in the project and 160 million rubles worth of equipment had already been ordered and that it would therefore be economically irrational to cancel the project. The deputy director of the Kurchatov Institute in Moscow also came out in favor of continuation, arguing that the energy produced was absolutely critical to the northwest's energy system.

The majority of speakers, however, came out against continuation of the project. The director of the Lithuanian Geological Work Association, Eduardas Banaitas, claimed that the designers had failed to take into account the earthquake potential in the area and had also planned poorly when locating the station on a groundwater zone used by Latvia, Belarussia, and Lithuania. The director of the Lithuanian Institute of Physical-Technical Energy Problems, Jurgis Vilemas, complained of incomplete information on the project. Deputy director of the Lithuanian Botany Institute, Romas Pakalnis, focused his complaints on the damaging effects of Ignalina on the marine life of Lake Druksai. Finally, president of the Lithuanian Academy of Sciences, Juras Požela, recommended canceling reactor #3 and using the allocated funds to build containment for the first two reactors. His proposal was also supported by vice president of the Lithuanian Academy of Sciences, Algirdas Žukauskas.

With the Lithuanian government on their side, the Lithuanian specialists finally won their long drawn out battle. After intense debate, a decision was reached to freeze construction of reactor #3 and appoint a special scientific commission to review its design and to decide its final fate. The decision was published in *Komjaunimo tiesa* on August 27.[30] The newspaper article focused on the tensions between the Lithuanian government and Moscow and explicitly asked who should be deciding such matters: Lithuania or Moscow? The article also noted the tension that had arisen between the Lithuanian government and the directorship of the Ignalina AES as a result of this conflict.

While this decision was certainly a victory for anti-nuclear activists, it was not viewed as the end of the battle. First, many people did not believe that construction of reactor #3 had actually been halted and rumors abounded of continued work at the site. Second, the final fate of reactor #3 had yet to be decided. And finally, the public was still concerned about the hazards of the reactors already in operation at Ignalina. During the summer of 1988, a series of fires at the station had lead to an upsurge in popular opposition to the continued operation of the station in general.[31]

Thus, popular opposition to Ignalina did not diminish with the roundtable decision to freeze construction of reactor #3. On September 7, Sajudis expressed its skepticism toward the actual cancellation of reactor #3 and condemned the continued operation of the first two reactors by sending appeals to Gorbachev, the International Atomic Energy Agency (IAEA) and the United Nations. The appeals requested that an international inspection team be sent to Ignalina to ensure the safety of reactors #1 and #2, and that reactor #3 be permanently canceled. The Sajudis appeal claimed to be based on overwhelming popular support for their concerns with Ignalina's safety, as demonstrated by 287,000 signatures on petitions opposing the station.

Shortly thereafter, the initiative group of Sajudis called on Lithuanians to participate in a blockade of the Ignalina station during the weekend of August 16–18. The planned meeting at Ignalina was in protest of the poor safety standards at the station and the construction of reactor #3, and was organized by Zigmas Vaišvila and Alvydas Medalinskas, both original members of the Žemyna club. While the protest was planned by Sajudis, it was also supported by the more radical nationalist organization, the Lithuanian Freedom League, which had already, on numerous occasions, stated its opposition to Ignalina.[32]

The meeting, however, ran into problems when local authorities refused to grant permission for a mass rally during that weekend on the grounds that the application was submitted late. Notification of the denial of permission was published in the republic press, along with warnings against attending this forbidden meeting. In fact, the Lithuanian and local authorities seem to have reacted quite negatively to the proposal for a mass meeting at the station, perhaps fearing the exacerbation of tensions between Lithuanian protestors and the predominantly Russian residents of the region.

Sajudis activists, however, decided to go ahead with the meeting, arguing that it had already been announced and it was too late to stop the flow of people to the area. In a blatant failure of glasnost, however, their decision to hold the meeting despite government opposition was blocked from appearing in any official publication. Sajudis members bitterly complained that the government used every organ in its control to publicize the cancellation of the meeting, while Sajudis was denied all access to the media.

Determination to hold the meeting was further fueled by a telegram Zigmas Vaišvila received from the International Atomic Energy Agency on September 15. The head of the IAEA, Hans Blix, informed Vaišvila that the safe operation of Ignalina AES was the responsibility of the USSR, not the IAEA, and the IAEA would only step in to review Ignalina's safety at the specific request of the USSR government. This information was quickly spread to the Lithuanian public in a television broadcast by Brazauskas that same day.

Sajudis organizers, however, made every attempt to avoid any kind of confrontation between protestors and authorities at the weekend demonstration. Instead of holding a "meeting," with speeches by anti-nuclear and national activists, Sajudis claimed that the weekend gathering was merely a chance for like-minded people to enjoy a few days of camping, theater, and musical entertainment.

Though Sajudis organizers feared a confrontation with authorities, the protest rally in fact went off without a hitch. While the authorities secured a strong police presence at the rally, the police were very careful to avoid confrontation and avoid interfering with the protest activities. On Sajudis's side, the organizers decided to cancel the religious service, which had been scheduled for Saturday night, on the grounds that it might bear too close a resemblance to a "meeting" and trigger a negative reaction from police and authorities. Thus, instead of the planned protest and blockade of the station, Lithuanians gathered to camp, plant trees and clean up the area, and enjoy plays, concerts and traditional dance performances. In their main action of the weekend, participants formed a human chain around the station, which they called the "ring of life."

While activists have claimed that the rally was attended by 50–100,000 people, Western observers have placed the number at closer to 20,000. This rally, however, is thought by many to have marked a real turning point in the psychological outlook of the population. While many had feared the consequences of holding a forbidden meeting, those who attended learned that the authorities would not interfere. Furthermore, having just achieved a significant victory on the Ignalina question at the August 25 roundtable, many people attending the rally were in an almost celebratory mood. People spoke of their sense of empowerment relative to Moscow and of a new freedom to play an active role in defining the future of their republic. This sense of achievement was further reinforced by an announcement from the Soviet government's official information agency, TASS, on Saturday, September 17, that the USSR government had agreed to officially invite the IAEA to review the safety of the Ignalina station.

The rally was also a chance to further solidify a sense of Lithuanian national identity. The Ignalina issue encapsulated the entire debate over who ruled Lithuania—the Lithuanians or Moscow—and Lithuania seemed to be winning the contest. The plays, music, and dance that entertained the protestors reflected a strong Lithuanian national element, and in many ways the weekend's activities were a celebration of the survival of the Lithuanian nation. By all accounts, the people who came from all over Lithuania to attend the rally were almost exclusively Lithuanian. It is also worth noting that the local residents from Ignalina and Sniečkus, who were almost entirely Russian in composition, did not support the rally and were apparently quite angered by the wave of Lithuanians who threatened their livelihood and security in Lithuania.[33]

The "ring-of-life" rally represented the culmination of the entire anti-Ignalina protest movement and, in fact, the final mass action that would be taken on this question. While the fate of the third reactor had yet to be conclusively resolved, and the question of the safety of the two reactors in operation had barely been addressed, both Sajudis and the public seemed to lose interest in the Ignalina question after the September rally. In fact, the growing authority of Sajudis relative to the Communist Party and the Lithuanian government during the summer and fall of 1988 allowed Sajudis to move beyond safe, environmental issues to more radical and political demands. Rather than focusing their attention on a surrogate issue, such as the Ignalina nuclear power station, Sajudis activists began to realize that they could now openly confront the real sources of their political grievances. Thus, Sajudis left the Ignalina issue behind and moved on to overtly political issues which were of much greater relevance to the question of republic sovereignty and independence.

The Final Chapter

During the fall of 1988, the scientific commission established as a result of the August 25 roundtable began its work investigating various aspects of the expansion project. Five working groups were set up to look at questions involving the seismicity, engineering, hydrogeology, water supply, and environmental impact of the station. The working groups were made up of nuclear specialists and representatives from the relevant USSR ministries. Representatives of Moscow far outweighed Lithuanians in each of the working groups.

In late 1988, the entire commission met to discuss the results of their research. According to several reports, three of the working groups favored continuing construction with certain design modifications, while two supported a pause in construction to address outstanding safety questions. None, however, called for the cancellation of the project. Following this session, the Lithuanian commission members gathered to formulate their own conclusions. Academician Vilemas, director of the Institute of the Physical-Technical Energy Problems and Professor Juodkazis of the University of Vilnius rushed to put together their own report and were able to present it to the chairman of the commission, vice president of the USSR Academy of Sciences, Academician Frolov, the following day.

A series of intense meetings between Frolov, the Lithuanian contingent, and other members of the commission followed. The Lithuanian protocol, as it became known, was finally signed by Frolov several days later, and became the official version of the commission's conclusions. It called for the cancellation of reactor #3 on the grounds of inadequate cooling facilities and for further investigation of the safety conditions at Ignalina in general. After active lobbying amongst the commission, approximately two-thirds of the members finally signed the protocol. The last third, however, remained adamantly opposed to its conclusions. The report was then sent on to the USSR Council of Ministers for official confirmation.

As relations between Lithuania and Moscow became more complex, however, and political turmoil swept through the republic, the Ignalina question was apparently forgotten. Construction at Ignalina was halted, but no final decision on the fate of reactor #3 was ever issued by the USSR Council of Ministers. In addition, the public seemed to have totally lost interest in the question. After September 1988, the press was devoid of articles on this question, and no more mass rallies were held.

The environmental movement that grew out of the activities of the spring and summer of 1988 blossomed and then faded quickly, as people turned their attention in new directions. An umbrella group, the Green Movement, was formed in October 1988 to unify the ecological movement, but its influence and membership diminished rapidly throughout the 1989–91 period, almost inversely with the growth of Sajudis. Later, the Green Party was formed with the goal of uniting ecological and political demands. By 1991, however, only a handful of members existed.

Thus, it seems apparent that the environmental and anti-nuclear movements were largely a surrogate that allowed intellectuals to voice their frustration with Moscow's colonial treatment of Lithuania or a tool which was used to activate a long-dormant society into public and political life. The commitment of its members to anti-nuclear or environmental causes, however, appears to have been marginal, leaving such movements extremely weak once opportunities for genuine political activism emerged.

The electoral victories of Sajudis in 1989 and 1990 and the achievement of genuine sovereignty over the 1990–91 period brought dramatic changes in attitudes toward nuclear power in Lithuania. As a result of these elections, leading anti-nuclear activists rose to positions of power in the Lithuanian government. Zigmas Vaišvila, the movement's most vocal and active member, became vice premier of Lithuania, while Alvydas Medalinskas, one of the key organizers of the "ring of life" and other anti-nuclear actions, became a member of the Lithuanian parliament and personal advisor to President Vytautas Landsbergis.

From this vantage point, the nuclear power issue took on a new meaning. This was especially true after the Lithuanian declaration of independence in March 1990. The declaration of complete independence from the USSR led to a strongly negative reaction from Moscow, which included a blockade of fuel into the republic. Suddenly, the Lithuanians found themselves dependent on their own sources of energy for heat and electricity. The importance of the Ignalina nuclear power station for the survival of Lithuania as an independent and self-sufficient state became all too apparent. On several occasions during the spring, technical problems at the Ignalina station forced the temporary shutdown of one or both of its reactors. The shutdowns triggered near crises for the Lithuanians and further reinforced the importance of the station for an independent Lithuania.

What is most intriguing is that the very activists that led the anti-Ignalina movement in 1988, suddenly began to shift to the other side of the issue. Not only did they fail to speak out against Ignalina after their elections, but they even went so far as to begin investigating the possibility of expanding the station. During the spring of 1990, a special commission was established within the new Lithuanian parliament to consider the possibility of expanding Ignalina and to address the question of what should be done about the strong Russian presence in Sniečkus and Ignalina.[35]

Both Vaišvila and Medalinskas were involved in the work of this commission, and both admitted at that time that they no longer harbored strong anti-nuclear sentiments. Interviews with both revealed that they viewed their participation in the anti-Ignalina campaign as a component in their rebellion against Moscow and their distrust of Moscow's ability to guard over Lithuanian interests and safety. If in the future a third reactor were to be built with help from Western nuclear specialists, both of these former activists said they would favor the project. Both also viewed popular opposition to Ignalina in 1988 as part of the struggle for independence, and claimed that reinitiation of construction at Ignalina would be unlikely to trigger a strong response from the population.

As of late 1993, the question of the reinitiation of the expansion project had yet to be resolved by the parliamentary group. By the end of that year, however, lack of funding as well as opposition to the expansion of the Ignalina station from neighboring Scandinavian countries appears to have derailed the possibility of renewed construction for the immediate future. In December 1993, the European Bank for Reconstruction and Development (EBRD) approved a substantial grant for upgrading the safety of the Ignalina AES. This grant, and the terms attached to it, were accepted by the Lithuanian government on February 10, 1994. Pushed heavily by the Swedish government, the EBRD grant provides for extensive safety upgrades in exchange for Lithuania's promise not to extend the life of the Ignalina station past the lifetime of the two reactors currently in operation (i.e., around the year 2010).[36] The grant also provides assistance to improve Lithuania's overall energy efficiency and to decrease its dependence on the Ignalina nuclear power station. Thus, interestingly enough, it was the safety and environmental concerns of neighboring Scandinavian and European governments and their populations, frightened by the threat of another Chernobyl, that ultimately prevented the expansion of the Ignalina station, rather than popular opposition from the population of Lithuania.

Anti-nuclear Mobilization in Armenia: *A Similar Trajectory*

Anti-nuclear mobilization in Armenia followed a path parallel to that observed in the Lithuanian case.[37] As in Lithuania, the movement was initiated by intellectuals whose deeper motives revolved around the national rebirth of Armenia. Furthermore, the surrogacy phenomenon was particularly pronounced in Armenia, with the anti-nuclear and environmental movements lasting only a few weeks before being supplanted by nationalist activism.

In Armenia, the official leadership proved very tolerant toward independent public activities as early as 1987. The first popular protests to emerge in Armenia were those opposing the Medzamor AES and the Nairit Chemical Factory, both located in close proximity to the capital city of Yerevan.[38] Within weeks of these environmental protests, however, the same activists who initiated them, emboldened by their success and the absence of government retaliation, moved on to state their true demands. Thus, environmental and anti-nuclear demands were heard for only a short period, before all attention was turned to the nationalist Nagorno-

Karabakh issue. Environmental activists quickly became the core for the initiative Nagorno-Karabakh Committee. Soon thereafter, interest in the nuclear power issue and environmental questions withered significantly.

Because the Armenian nationalists encountered a relatively hospitable environment for voicing their demands and mobilizing people to the nationalist cause, there was no need for the activists to continue to hide behind anti-nuclear and environmental slogans. In addition, Armenia's high percentage of ethnic Armenians as well as its long national history and well-defined cultural identity, provided a strong and coherent national identity around which people could quickly mobilize.[39] Thus, demands for national sovereignty and the unification of Armenian lands sprang forth almost immediately after the opportunity arose.

While Armenia's sole nuclear power station was quickly closed in response to public opposition,[40] it was not long before popular attitudes toward the station began to change. During the late 1980s, the conflict between Armenia and Azerbaijan resulted in a virtual blockade of Armenia, and the republic suddenly found itself dependent on its own supplies of energy to provide heat, electricity, and power for industrial manufacturing. As in Lithuania, increasing sovereignty in Armenia led to widespread reassessment of the significance of the republic's nuclear power station. Suddenly, possession of a nuclear power station became linked to survival as an independent nation, and the public began to view the Medzamor station in a much more positive light.

Following the breakup of the Soviet Union, Armenia's parliament voted to relinquish its role in determining the fate of the Medzamor station and to permit President Ter-Petrosyan to decide whether or not to reopen the station. Beginning in 1991, specialists from Moscow were brought to Armenia to assess the possibility of reopening the closed station. Later, Western specialists were brought in to make recommendations as to what should be done to bring the shutdown reactors back on line. By 1993, President Ter-Petrosyan had clearly come out in support of reopening the Medzamor station and was concerned only with the question of financing the project.[41] On January 26, 1995, the Russian Duma voted to provide 60 billion rubles to Armenia to finance the reopening of the Medzamor station.[42] The station went back in operation in October of 1995.[43] Throughout this period of debate and reconstruction, there have been no reports of popular protest and no significant public opposition to plans to reopen the station.

Thus, in both Lithuania and Armenia, early opportunities for mobilization on radical platforms and a strong, unified sense of national identity facilitated the emergence of a powerful nationalist movement. In both cases, opposition to nuclear power appeared to be little more than a stepping stone toward the creation of a national-sovereignty movement, and the anti-nuclear movements faded quickly as nationalist movements grew.

CHAPTER 3

Ukraine: Civic or Ethnic Mobilization?

Introduction

While the anti-nuclear cause was used by Lithuanian intellectual elites as a tool to mobilize society in support of independence, the linkage between anti-nuclear activism and nationalism in Ukraine was much less direct. Rather than being a surrogate for another cause, in Ukraine concerns about nuclear power safety were very powerful and very real. The Chernobyl experience had traumatized Ukrainian society in a way not paralleled in Lithuania. The Ukrainians had seen firsthand the results of Moscow's incompetence in the nuclear power sector, and the nuclear power issue thus took on a special resonance. In addition, the Ukrainian national identity proved to be unexpectedly weak, and nationalism turned out to be a much less effective mobilizing force in Ukraine than it had been in the Baltic republics. Thus, the relationship between the anti-nuclear movement and the nationalist cause in Ukraine was a complex one; the causes were intertwined, but neither was a surrogate for the other.

While the Lithuanians had only a single target on which to concentrate their energies in their anti-nuclear crusade, the Ukrainians were confronted with a much more ominous situation in the nuclear power sector. In the late 1970s, a decision was made in Moscow to dramatically expand and develop nuclear power facilities in European USSR. In particular, the Soviet leadership decided to construct a large number of reactors in Ukraine and the western portion of the Russian Federation. The first nuclear power reactor in Ukraine went into operation in 1977 at the Chernobyl AES. By early 1985, there were already ten reactors in operation in Ukraine,[1] and nuclear power made up approximately 20 percent of Ukraine's power capacity.[2] While the operating nuclear power stations in Ukraine represented a total capacity of 8,800 MW in 1985, the Twelfth Five-Year Plan (1986–90) called for rapid construction of new facilities leading to a total capacity of 32,880 MW by 1990.[3] This meant that the post-Chernobyl period was to be a time of frenzied construction and unprecedented growth in the nuclear power sector in Ukraine.

Despite the powerful impressions created by the Chernobyl disaster, anti-nuclear activism was slow to emerge in Ukraine. While the Lithuanian movement reached its peak in the summer of 1988, the most active period for the Ukrainian movement was 1989–90. Ukraine's apparent passivity during the first several years following the Chernobyl disaster can be attributed to a combination of factors. Because the hard-line first secretary of the Communist Party of Ukraine (CPU), V. V. Shcherbitsky, was able to maintain his hold on the Ukrainian government until September 1989, the introduction of perestroika was significantly delayed in Ukraine.

Secondly, unlike in Lithuania, physical scientists within the Ukrainian Academy of Sciences did not constitute a base of intellectual opposition to the regime, and the Academy did not provide a safe haven for the budding movement. A network of scientists opposing Moscow's nuclear power program did not exist prior to the late 1980s; thus there was no scientific elite ready to quickly step in and assist in mobilizing society against nuclear power in Ukraine.

The most critical factor, however, in differentiating the Lithuanian and Ukrainian anti-nuclear movements was the strength and nature of the republics' national identities. Whereas Lithuanian intellectuals were unified in their understanding of what constituted the Lithuanian nation and found it relatively easy to mobilize a society which was 80 percent ethnic Lithuanians in support of their cause, the Ukrainians found themselves in quite a different situation. Because much of Ukraine had never experienced independent statehood and because the Russian and Ukrainian histories, languages, and cultures are so closely intertwined, Ukrainian intellectuals found it difficult to unite around a single national idea. Furthermore, because such a large proportion of the population was considered ethnic Russian and many people were quite simply confused about what it meant to be a Ukrainian, mobilizing a Ukrainian national movement was much more difficult. As a result, the national movement in Ukraine was much weaker and less focused than in Lithuania. Thus, the anti-nuclear and national movements grew up side by side, and while each adopted the other's cause into their own movement, the two movements never merged into one, and neither served as a surrogate for the other.

In this chapter, I will trace the emergence of the anti-nuclear and nationalist movements in Ukraine, and consider differences in the mobilizational patterns in Ukraine and Lithuania. An overview of Ukrainian mobilization will be presented, focusing in particular on the following questions: What role did intellectual elites play in anti-nuclear and national mobilization? How did the distribution and availability of mobilizational resources affect movement profiles and developmental patterns? How can we explain the relative passivity of the Ukrainian population, its reluctance to mobilize, and its lack of strong support for the nationalist cause? How was national identity defined in Ukraine? And finally, what explains the collapse of both the green and nationalist movements in 1991–92?

Chapter 4 will be devoted to a detailed study of one local anti-nuclear movement. In this chapter, I will discuss the rise and fall of the movement against the Khmelnitsky nuclear power station, and show how mobilization occurred at the local level, far from the republic capital. This in-depth look at a grassroots movement in Ukraine will provide much greater insight into the relationship between the intellectual elites organizing in Kiev and the grassroots activists who made up the bulk of the anti-nuclear mass movement.

The Emergence of Anti-nuclear Activism in Ukraine

Before the Disaster

Although a group of top-level scientists within the Lithuanian Academy of Sciences had long been concerned with Moscow's plans to build a gigantic nuclear power station in their republic, there is little indication that a parallel group existed within the Ukrainian Academy of Sciences. Interviews with both nuclear specialists and other scientists from the Ukrainian Academy of Sciences confirm that there was little discussion among Ukrainian scientists on the issue of nuclear power. The president of the USSR Academy of Sciences, A. Alexandrov, strongly supported the expansion of nuclear power in Ukraine, and most Ukrainian scientists acknowledge his extremely tight control over the entire nuclear sector in Ukraine.

One factor that may have impeded the growth of anti-nuclear sentiment among the scientific community in Ukraine was the fact that Ukraine had no academic facilities for training nuclear specialists. Thus, virtually all of Ukraine's nuclear scientists were trained in Moscow, and the majority of nuclear specialists working in Ukraine were not native Ukrainians. Ukraine also did not possess its own branch of the USSR Ministry of Atomic Energy and Industry nor its own Atomic Control Commission. Most scientists describe the relationship that dominated the nuclear power sector in Ukraine as a "colonial" one.

There is no sign that, prior to 1986, high-level scientists within the Ukrainian Academy of Sciences ever questioned or opposed Moscow's plans to dramatically expand nuclear power production in Ukraine in the 1980s and 1990s. There is no indication that a network of anti-nuclear scientists existed within the academy or that the academy provided any institutional assistance to anti-nuclear activists once the movement began to emerge in the late 1980s.

The situation among the general population prior to 1986 differed little from that of the scientific community. Due to the total lack of objective information on nuclear power and to the propaganda on the "absolute safety" of nuclear power stations which deluged the population, most people did not even consider the possible dangerous implications of the enormous nuclear power stations being constructed in all corners of their republic.

There were no opportunities for people to mobilize against nuclear power prior to 1985, and any attempts to speak out against nuclear power stations would have met with a vigorous reaction from the KGB. Yurii Shcherbak, a prominent writer and leading environmental activist of the perestroika period, insists however that the lack of opportunity for mobilization prior to 1985 was not a factor. According to Shcherbak, opposition to nuclear power simply did not exist among the general population prior to Chernobyl.[4]

The only group to take a significant interest in the environmental degradation of Ukraine prior to 1986 can be found within the writers' community. Thus, in the mid-1960s, Ukrainian writers began to take an interest in environmental questions. Small groups occasionally met to discuss their concerns about damage to the Dnieper River, air pollution, hydroelectric stations, and other kinds of environmental degradation in the republic. A few articles were even published in literary journals on environmental topics. The writers, however, were not particularly concerned with the nuclear power question. Writer Yurii Shcherbak has estimated that in the 1970s, a tiny group of perhaps three to five writers met to discuss their concerns about nuclear power, but this concern did not extend beyond a half dozen writers, and the issue was never a core concern for environmentally oriented writers.[5] No articles on nuclear power were even published in the literary press prior to 1986.

Chernobyl and the Birth of Perestroika in Ukraine

While the accident at the Chernobyl nuclear power station occurred in April 1986, society was slow to react to the catastrophe. Although a tiny portion of the cultural intelligentsia began to reassess their attitudes toward nuclear power in Ukraine during the first two years following the disaster, the bulk of society remained passive and apparently unconcerned about the nuclear power threat in their republic. They were slow to digest the new information on nuclear power safety and to begin questioning Moscow's competence to safely operate nuclear power stations in their republic. People were so accustomed to blindly accepting state propaganda on the wonders of nuclear power that it took time for them to begin to cast off their blinders and make their own judgments on this issue. Many people have described the two-year period following the accident as a period of shock; society was confronted with an unexpected and horrible reality which had to be slowly incorporated into their understanding of the world around them.

The process of reassessing longstanding beliefs about the "absolute safety" of nuclear power was also greatly hampered by the lack of genuine glasnost on the nuclear power issue. While Western commentators were quick to point to Soviet reports on the Chernobyl disaster as a sign of a new openness in the Soviet media, their praise was largely unwarranted. In fact, a detailed survey of the Soviet and Ukrainian press during the

1986–87 period indicates that information on the accident was still highly restricted and published reports were often intentionally falsified to obscure the true magnitude of the disaster.[6] While the high-circulation press permitted publication of articles dealing with the progress of the accident cleanup and investigation into its causes, no articles were published which questioned Moscow's competence to safely operate nuclear power stations or the government's plans to dramatically expand nuclear power facilities in Ukraine during the post-Chernobyl period. Thus, the bulk of society was not only denied crucial information on the disaster which might have altered their attitudes toward nuclear power, but they were also prevented from hearing the views of opponents of nuclear power. Intellectual elites opposing nuclear power in Ukraine were given no opportunity to present their views to society and to attempt to mobilize people in support of their cause.

Members of the intelligentsia, however, were beginning to reassess their own views on nuclear power during the early post-Chernobyl period. The writers' community, in particular, was quick to react to the disaster and began a campaign against what they viewed as a threat to the survival of their republic. Thus, in 1986–87, such leading writers as Oles Honchar began to focus their attention on this issue and attempt to publicize their concerns in society.[7] While the high-circulation press was closed to such politically dangerous viewpoints, the newspaper of the Ukrainian Writers' Union, *Literaturna Ukraina,* published a number of bold articles on nuclear power during 1986 and 87.[8] Although these articles were unlikely to have been viewed outside a narrow stratum of society, they did help promote awareness about the nuclear threat among the writers' community and other intellectual elites. Thus, it is not surprising that the writers were the first to take the bold step of organizing an independent association on this issue to begin the struggle against Moscow's unflinching commitment to make Ukraine the centerpiece of the USSR's nuclear power program.

Zelenii Svit: An Umbrella for Ukrainian Greens

Opposition to the continuation and expansion of nuclear power in Ukraine began in the Ukrainian Writers' Union. From early 1987, concerned writers began to gather and question plans for the continued development of nuclear power in their republic. While a number of well-known writers became involved in these discussions, the leading force in organizing this intellectual opposition into a movement was Dr. Yurii Shcherbak, a well-respected physician and writer.[9]

The first action taken to bring writers concerned with the nuclear power issue together was a conference held in Kiev in April 1987. The conference was organized by Shcherbak and hosted by the Ukrainian Writers' Union. Although the topic for the conference was the deterioration of Ukraine's environment in general, it was the first forum for openly discussing the nuclear power program in Ukraine and beginning to question the wisdom of its continuation. Writers from across Ukraine, most of whom had long been concerned with environmental questions, were invited to attend the conference. According to Shcherbak, these individuals constituted a network of intellectuals that he was already personally well-acquainted with; while intellectuals outside the writers' community may have also harbored concerns about nuclear power, these individuals were not known to the writers and thus were not included in the initial stages of movement organization.

Within the scientific community, opposition to the planned expansion of nuclear power in Ukraine was in fact growing. Such opposition first started to coalesce around the question of whether the almost-completed fifth and sixth reactors at Chernobyl AES should be permitted to go into operation. In March 1987, more than sixty scientists from various branches of both academic and industrial science met in Kiev to discuss the fate of these new reactors.[10] At the end of the meeting a vote was taken as to whether to recommend continuation of construction, and it was reported that only two scientists voted in favor of the recommendation.[11] Shortly thereafter, Moscow announced its decision to halt construction on these two reactors and institute a temporary freeze on plans to bring the reactors into operation in the near future.

During the remainder of 1987 other meetings of scientists and engineers were held to discuss the nuclear power situation in Ukraine. A particularly important scientific meeting was held on August 25, 1987, to consider the planned expansion of the existing nuclear facilities at Rovno, Khmelnitsky, and South Ukraine stations. At this meeting, scientists agreed that the maximum capacity of an individual nuclear facility should not exceed 4,000 MW, and thus recommended to the Ministry of

Atomic Energy that the expansion projects at the Rovno, Khmelnitsky, and South Ukraine AESs be halted.[12] During the fall of 1987, however, the Ministry of Atomic Energy reconfirmed its intention to continue all three expansions and was even able to bring at least one of the new reactors on line by the end of the year.[13]

While individuals within the scientific community clearly opposed the planned expansion of nuclear power in Ukraine, there is no indication that any steps were taken toward organizing this opposition into a unified movement. The only known exception to this tendency was a tiny group of power engineers within the Ukrainian Ministry of Power's scientific-production association, the Obshchestvo energetikov (Association of Power Workers). Within this association, a handful of power specialists were united in their opposition to the continued expansion of the Chernobyl AES and eager to play a more active role in slowing nuclear construction in Ukraine.[14]

In late 1987, the growing network of anti-nuclear writers and this tiny core of power specialists finally became acquainted with each other and began to work together as a single movement. In December 1987, *Literaturna Ukraina* and the Ukrainian Communist Party newspaper *Radyanska Ukraina* organized a conference entitled, "Scientific-Technological Progress and Morality," which brought together both writers and scientists to discuss ethical dilemmas associated with high technologies.[15] While this conference was apparently dominated by writers, members of the scientific community did participate. According to many reports, the main topic of discussion at this conference revolved around the question of the future of nuclear power in Ukraine. In their speeches, the writers made ample use of the scientists' concerns about specific nuclear expansion projects, thus demonstrating increased acquaintance and cooperation with the scientific faction.

Finally, in December 1987, parallel steps were taken toward creating an independent environmental association. In the Ukrainian Writers' Union, an environmental commission was formed, with Yurii Shcherbak as its chairman. Simultaneously, the writers met with other members of the intellectual community, in particular, concerned members of the Ukrainian Union of Cinematographers and the handful of power specialists concentrated in the Obshchestvo energetikov, and decided to form a more broadly based independent association, which they called Zelenii svit (Green World). The initial composition of Zelenii svit included no more than a few dozen intellectuals, but the very fact of its creation under such adverse conditions was remarkable.

While the Ukrainian environmental movement did not find shelter in the Academy of Sciences, as their Lithuanian counterparts had, they were not without their own protectors. First, the Ukrainian Writers' Union had long-provided a relatively safe haven for environmental activists within the writers' community. At the end of 1987, however, the fledgling movement received an even greater boost when the official Peace Committee (Komitet zashchiti mira) agreed to take Zelenii svit under its wing and provided the new group with meeting space, minimal funding and technical support, and a more protected official status. As usual, the godsend for the new movement was the result of personal connections; the chairman of the Peace Committee in 1987 was in fact Oles Honchar, a leading writer and one of the first to begin the anti-nuclear power publicity campaign in Ukraine. Thus, Honchar's personal sympathies for the new movement played a key role in the Peace Committee's decision to act as a sponsor for Zelenii svit.

During 1988, the writers continued to dominate the environmental movement, and little effort was made to appeal to members of society outside of a narrow stratum of intellectuals. In March 1988, Zelenii svit took its first step toward organizing itself into an all-republic organization. Writers from every corner of the republic were invited to attend a conference in Kiev on the environmental, and particularly nuclear-power, situation in Ukraine. Again Shcherbak was the principal force behind this conference, and writers were the main participants in the conference. Approximately 200 writers attended the meeting. Discussions were reported to be quite heated, and the final outcome of the conference was a set of resolutions demanding the eventual closure of the Chernobyl nuclear power station and a freeze on all new nuclear construction in Ukraine. Resolutions on other environmental questions were also adopted.

This conference, however, was more than simply an opportunity for writers to discuss environmental concerns. It was in fact the forum for expanding Zelenii svit from a merely local Kiev club to an all-Ukraine umbrella organization. During the meeting, concerned writers from across Ukraine began to discuss how to go about creating an all-republic environmental organization and how such a movement should be organized.

Following this conference, leading activists began to meet regularly in Kiev to work toward establishing an all-Ukraine organization. While the leadership of the movement continued to be based in Kiev, participants now included representatives from other regions, and the agreed-upon goal was to construct a movement which united all environmental forces in the republic.

During 1988 and early 1989, conditions were not auspicious for independent activities in Ukraine. Unlike in Moscow, Leningrad, and the Baltics, large rallies and demonstrations were still forbidden and independent activities were viewed with great suspicion. While Zelenii svit attempted to organize a memorial rally on the second anniversary of the Chernobyl disaster, April 26, 1988, their appeal for permission to hold such a public gathering was denied. Nevertheless, the activists took the bold step of holding the memorial demonstration illegally, and discovered to their amazement that few repercussions followed. The rally albeit small was another ground breaking step in the growing confidence of the new informal movements.

While some tentative independent activities were beginning to emerge in large cities such as Kiev and Lvov, activists in the countryside faced huge obstacles to mobilization and saw little improvement in their mobilizational opportunities relative to the pre-perestroika period. Nevertheless, in 1988 small environmental clubs began to form across Ukraine in regions where environmental threats were viewed as both immediate and severe; the clubs emerged to fight huge industrial threats in people's immediate neighborhood. Given the experience of Chernobyl, the construction of new nuclear facilities proved to be the most potent cause around which these new local environmental groups were mobilized.[16] Thus, in 1988, local groups began to emerge to oppose the nuclear-construction projects at the Rovno, Khmelnitsky, South Ukraine, Crimean, and Chigirin AESs.[17]

In September 1989, with the final ouster of hard-line communist, V. V. Shcherbitsky, perestroika finally began to make its way into Ukrainian life, and independent activities began to blossom. The period from 1989 to 1990 was thus the golden age for Zelenii svit. Press restrictions on articles concerning nuclear power diminished considerably; repression and harassment of independent activists in both city and countryside fell (albeit, not equally in all regions); and opportunities for movement mobilization expanded rapidly.

On July 27, 1989, Zelenii svit became the first independent organization to officially register its existence in Ukraine.[18] The group was registered as an all-republic organization, and its request for the right to register local chapters without the approval of local authorities was unexpectedly approved. As a result, new chapters only had to apply to the Zelenii svit leadership in Kiev in order to receive the rights of a registered "social organization." Just what these rights included, however, was never clear. The most recent specification of the rights of social organizations was found in the old Stalin constitution of 1936, which had never actually been implemented. Registration did, however, provide the local activists with some degree of protection from meddling authorities and apparently conferred the right to hold bank accounts, to petition for permission to hold demonstrations, and to hold regular gatherings of members.

During 1989, members of the scientific community began to be drawn into the Zelenii svit organization. Though nuclear specialists generally showed little sympathy for the organization, scientists from other sectors were more open to the Zelenii svit message. By late 1989, scientific participation in Zelenii svit was up substantially, and sympathy within the Ukrainian Academy of Sciences for Zelenii svit's anti-nuclear platform was growing.

Throughout 1989, anti-nuclear movements and other environmental clubs blossomed across Ukraine, and finally, in October 1989 the founding congress of the Ukrainian environmental movement was held in Kiev. At this congress, Zelenii svit was established as an umbrella for the multitude of local environmental and anti-nuclear clubs that had sprung up across the republic since the end of 1988. Representatives from almost every oblast of Ukraine attended the founding congress and voted to confirm the movement statutes, the program, and a set of fourteen environmental resolutions. The program and the resolutions included demands that all new nuclear projects in Ukraine be abandoned, that Chernobyl AES be closed, and that a Ukrainian department of the USSR Ministry of Atomic Power be established to supervise Ukraine's nuclear program.

In organizing Zelenii svit, the founders rejected the principal of "democratic centralism." Rather than creating a bolshevik-type organization with strict vertical lines of authority, the leading activists agreed that Zelenii svit should be a horizontal, grassroots organization. The leadership was there to advise local activists and to assist them, but no orders could be given from above. Local activists were free to pursue whatever strategy and tactics they preferred. A council of approximately 100 members, composed of the leaders of local clubs, was to meet monthly to coordinate activities of local chapters. At the very top of the movement, an elected secretariat of about a dozen top activists was to direct movement activities in Kiev. At the founding congress, Yurii Shcherbak was elected chairman of the movement.

Following the congress, the environmental movement blossomed rapidly across Ukraine. By late 1989, conditions for informal activities were finally beginning to improve, and clubs of all kinds were springing up. The first mass demonstration in Kiev numbering in the tens of thousands was held in November 1989. The demonstration was sponsored by several informal organizations, including Zelenii svit, and spanned a wide variety of political, environmental, and anti-nuclear demands. Ukrainian society was finally coming to life, after years of silence and passivity. More demonstrations were held in Kiev in early 1990, with an environmental protest on January 28 and a green contingent marching in the annual May Day demonstration. This was the first time the greens had participated in a May Day demonstration, and the government authorities reacted quite negatively. Participants were hassled by the police, and one person carrying an anti-Chernobyl poster was even fired from his job immediately thereafter.[19]

By the spring of 1990, approximately 300 local clubs had registered under the Zelenii svit umbrella.[20] Journalists have estimated that this constituted about 90 percent of the informal environmental organizations existing at that time in Ukraine.[21] The movement was strongest in the western and southern regions of Ukraine and around nuclear power stations. It remained quite weak in the Russian-dominated eastern territory of the republic.

During 1989, Zelenii svit focused its attention on decision makers in Moscow and attempted to convince both Gorbachev and the USSR Congress of Peoples' Deputies of the need to curtail Ukraine's nuclear power program. Local clubs collected signatures against the construction of nuclear power facilities in their regions, and these anti-nuclear petitions began accumulating in Zelenii svit's Kiev center. During 1989, tens of thousands of signatures were collected against the Khmelnitsky, Chigirin, South Ukraine, Chernobyl, and Crimean AESs. According to several members of the Zelenii svit secretariat, over 250,000 anti-nuclear signatures were collected by November 1989. During the year, these petitions were steadily forwarded directly to Mikhail Gorbachev and the USSR Congress of Peoples' Deputies.

During 1990, however, movement tactics changed substantially. Though collecting signatures against Soviet policy on nuclear power seemed daring and dramatic in 1989, by 1990 activists were pushing the bounds of permissible activity even further. Disruptive tactics, such as strikes, protests, picketing and station blockades, began to dominate the action repertoire of local activists. As people learned that the authorities were unwilling to deal severely with the perpetrators of such actions, disruptive tactics spread like wildfire. Much like in Lithuania, local activists constantly tested the waters and watched to see how far the limits on independent activities could be pushed.

While Zelenii svit took the lead in organizing the anti-nuclear movement in Ukraine, it was not alone in opposing Moscow's nuclear power program. The budding national-sovereignty movement was also quick to adopt the cause and in fact played a key role in mobilizing society against nuclear power stations in Ukraine. Unlike in Lithuania, however, the two movements remained separate; the goals of the two movements were intertwined but not congruent. The relationship between the two movements incorporated both cooperative and competitive aspects, and the logic behind their dual existence was not always obvious.

Rukh: Political Mobilization in Ukraine

While the political climate of Ukraine in the late 1980s forced intellectuals to devote their energies to causes that were not viewed as overtly political or threatening by the old regime, independent political movements finally began to emerge in early 1989. Many political activists note that while their participation in informal groups began with fairly innocuous causes—such as the Stalinist legacy, Ukrainian cultural and literary revival, and environmentalism—they quickly became politicized as they encountered the hostility and resistance of the old regime to any kind of independent activity. Thus, by late 1988, a network of potential political activists was beginning to emerge in Ukraine. Once again, the core of this network was located within the Ukrainian writers' community.

While initially two parallel initiative groups grew up within the Kiev branch of the Writers' Union, one within the party organization and the other independent, the two initiative groups eventually merged in the winter of 1988–89. In January 1989, a draft program for the new "Movement for Perestroika" was composed within the Writers' Union.[22] Meanwhile, during the spring of 1989, intellectuals outside the writers' community (for example, in the Institute of Philosophy) were beginning to form their own initiative groups, and in March 1989 a Coordination Council was established to bring together the various initiative groups within the intellectual community.

While the founders of the new movement, commonly known as "Rukh," were careful to stress their support for Gorbachev's program of perestroika and to reject all suggestions that the movement might represent a political opposition group, the Central Committee of the Communist Party of Ukraine reacted very negatively to Rukh's formation. An extremely harsh and vitriolic press campaign was initiated against the founders of Rukh, accusing them of being antistate, antiparty, anti-Semitic, and worse.[23] The party newspapers were filled with letters protesting the formation of this subversive group.[24] Since many of the founders were party members, some found themselves officially censured by the party, while others were expelled from the party. At a meeting of the general assembly of the Ukrainian Academy of Sciences in March 1989, the Academy officially condemned the formation of Rukh and warned its members to stay away.[25]

With the retirement of hard-line First Secretary Shcherbitsky in September 1989, the attacks against Rukh finally began to simmer down. On September 8–10, the organization was finally able to hold its all-Ukraine founding congress. At the founding congress, it was constantly repeated that the movement was not antistate, antiparty, or pro-independence for Ukraine. The founders of Rukh initially attempted to define themselves as an apolitical group whose only goal was to help promote the policies of Gorbachev in Ukraine. It was not until 1990 that the movement began to become overtly politicized and to move in the direction of opposition to the party and independence for Ukraine.

Rukh's role in the environmental and anti-nuclear movements was, at least officially, quite minimal. While environmentalism was one component of the Rukh program, it was not central to the founders' concerns and received only cursory treatment in the program. An environmental section of Rukh was formed, but its activities were minimal.

Despite the fact that Rukh's orientation was not primarily focused on environmental affairs, the movement did play a key role in the mobilization of the anti-nuclear movement. As Rukh began to radicalize and demands for sovereignty began to creep into the movement, the overlap between anti-nuclear demands and demands for more decision-making rights became quite obvious to all. Thus, it was only natural for the increasingly anti-Moscow and independence-minded movement to assist anti-nuclear protestors. As in Lithuania, Moscow's nuclear policies in Ukraine were presented as threatening the destruction of Ukraine, and thus they were not only an environmental but also a national concern.

It is important to note, however, that although Rukh supported the anti-nuclear movement, it rarely went as far as their Lithuanian neighbors did, to accuse Moscow of fomenting a policy of "genocide" against the Ukrainian people. In fact, the Rukh movement took deliberate steps to define itself in civic rather than ethnic terms and to avoid aggravating any kind of anti-Russian sentiment in the population. The Rukh leadership repeatedly insisted that the movement was for the sovereignty of the republic of Ukraine, based on equal rights for all residents, regardless of ethnic identity. While Rukh branches in West Ukraine occasionally seemed to shift in an ethnic direction, the overall orientation of the movement was explicitly civic, and leaders of the republic organization went to great extremes to convince the population that their goal was improved living conditions for all peoples in Ukraine.

It is also important to note that although Rukh attempted to duplicate the successes of the Baltic peoples' fronts, national-democratic mobilization in Ukraine was much weaker than in the Baltic republics. While mass rallies did occur in Ukraine in 1990–91, they never reached the fever pitch of the Baltic mobilizations, and support for an independent Ukraine was only beginning to blossom in 1991 when independence was suddenly tossed into their laps.

Thus, national identity was a much weaker mobilizing force in Ukraine than it had been in Lithuania. Rather than anti-nuclear protest acting as a surrogate for nationalist aspirations, in Ukraine the two demands became intertwined and complementary. In fact, the environmental and anti-nuclear movements served to reawaken a long-dormant society and remind them of their national heritage. Calls to protect the Ukrainian lands and people from the nuclear threat were important to the resuscitation of a sense of Ukrainian national identity. In Ukraine, environmental and nationalist goals were mutually reinforcing; it might be more accurate to think of the anti-nuclear movement as a catalyst for nationalism rather than a surrogate.

1990: A Change in Strategy

While anti-nuclear activists relied primarily on mass rallies and petition drives during 1989, the introduction of multicandidate local elections in 1990 brought a new element to movement strategy. In contrast to the Lithuanian case, the local elections in Ukraine did not bring the greens and democrats to power. Because perestroika was introduced in Ukraine much later than in the Baltic republics, the old communist apparatus remained quite strong in 1990 and was able to block the nominations of Rukh and Zelenii svit candidates. Thus, candidates nominated by Rukh, Zelenii svit, and other informal associations were uniformly rejected by the local electoral commissions.

The ban on candidates from informal organizations met with outrage among Rukh and Zelenii svit members, but they were unable to force the old apparatus to open the electoral process to the new informal groups. In protest, Yurii Shcherbak withdrew his candidacy to the Ukrainian parliament. In addition, Rukh and Zelenii svit jointly sponsored a rally in Kiev to demand registration of their candidates. The rally was attended by tens of thousands of people but had little impact on the decisions of the electoral commissions.

In response to this dilemma, Rukh leaders acted quickly to encourage their members to seek alternative nominating organizations. Thus, many Rukh candidates finally made it onto the ballot as nominees from local factories, institutes, and other official organizations. The Zelenii svit leadership, however, continued to protest the unfair registration procedures and did not take steps to register their candidates through other organizations until the very last moment. This was widely acknowledged later to have been a major error in strategy. Nonetheless, a handful of Zelenii svit members were registered as candidates, and six members of the organization were eventually elected to the Ukrainian parliament. The success level of the green movement at the local level was somewhat higher, but genuine environmentalists still made up just a tiny contingent in most local soviets. While Rukh did comparatively well in local elections in West Ukraine, the Ukrainian parliament was dominated by conservatives (commonly known as the "group of 239"), and the new democrats were in the minority.

That there were few Zelenii svit candidates does not mean that the nuclear power issue did not play a role in local elections. In fact, as will be discussed in more detail in the following chapter, almost all candidates running in regions located near controversial nuclear power stations professed their vehement opposition to nuclear power. Thus, while the new local soviets were not composed primarily of genuine environmentalists, their members had almost unanimously opposed local nuclear power stations during their electoral campaigns. Thus, the new local soviets were at least superficially strongly anti-nuclear in their composition.

Following the local elections, the new city and oblast soviets located near nuclear power stations found themselves under great pressure to fulfill their campaign promises to cancel the new nuclear facilities in their regions. The first oblast soviet to make such a decision was in Khmelnitsky. Others, however, quickly followed. Thus, the local soviets suddenly claimed for themselves the right to make decisions regarding major construction projects in their regions. Rather than appealing to Gorbachev and the USSR authorities, local soviets took matters into their own hands and canceled the projects themselves. This represented a dramatic change in established decision-making procedures in the nuclear power sector.

While these local decisions were significant in their attempt to claim new local powers and privileges, they in fact had little impact on nuclear power policy. Because the construction funds flowed from Moscow and the AES administrations viewed the USSR government as their ultimate authority, construction continued at all of these stations. The population had won a hollow victory at best.

In August 1990, however, the Ukrainian parliament decreed a five-year moratorium on the construction of new nuclear projects in the republic.[26] While the wording of the decree was extremely vague and left open the question of what should be done with partially completed stations, most anti-nuclear activists saw the moratorium as a complete victory for their cause. While construction continued at most of the nuclear power stations in the republic, anti-nuclear activism quickly withered away after this symbolic watershed.

In fact, the environmental movement in general was quick to disappear after the moratorium decision. While leading activists in Kiev had counted their membership in terms of the numbers of signatures on environmental and anti-nuclear petitions and had boasted of perhaps 200,000 members in 1989, they suddenly recognized that their movement was much weaker than they had thought. In fact, only a handful of activists remained scattered across the republic, and they all complained that the population was no longer interested in environmental or nuclear matters. It was no longer possible to mobilize a mass rally on environmental issues.

After August 1991, the tiny core of anti-nuclear activist left trying to keep the movement alive saw their situation deteriorate significantly. The introduction of independence in Ukraine suddenly meant that nuclear power no longer represented Moscow's dominance in Ukraine; instead, it came to symbolize Ukraine's potential to sustain itself as an independent and self-sufficient country. Thus, after the dissolution of the USSR, attitudes toward nuclear power changed dramatically—both among the general population and the political elite. In December 1993, the Ukrainian parliament voted to overturn its moratorium on nuclear construction.[27] By all reports, this decision was not hotly debated among parliamentarians and triggered little reaction from the general population. Since that time, construction has resumed on a number of previously suspended projects, including the Khmelnitsky, Zaporozhoye, and Rovno AESs, and new reactors have already come on line at several stations.

Interviews with leading environmental and anti-nuclear activists in Ukraine after 1991 revealed widespread disappointment and disillusionment with the fate of the movement. Activists openly admitted to a lack of public interest in the anti-nuclear cause and the virtual impossibility of mobilizing mass actions against nuclear power at this time. Interestingly enough, however, by 1995 new signs of anti-nuclear activity were beginning to be apparent. With expanding opportunities for foreign organizations to establish branches in the former Soviet Union and to channel aid to preferred causes, Western interest in curtailing nuclear expansion in the former USSR became a factor in organizing and funding anti-nuclear activities. Numerous Western organizations are currently supporting Ukrainian groups who profess to oppose Ukraine's nuclear power program. In addition, such anti-nuclear giants as International Greenpeace have established branches in Kiev and are attempting to promote the anti-nuclear crusade. As yet, however, there is little sign that this Western aid has had any impact on mass sentiment in Ukraine. Opposition to nuclear power currently remains a marginal phenomenon and the anti-nuclear platform has ceased to be a rallying platform for the vast majority of Ukraine's society.

Conclusions

If the anti-nuclear movement in Ukraine was based on genuine fears about nuclear power and was not simply a surrogate for nationalist aspirations, why did it wither away so quickly in 1991 and 1992? There are several possible explanations for this phenomenon. First, while not being a surrogate for nationalism, the anti-nuclear movement certainly incorporated an anti-Moscow element which was no longer needed when Moscow's control over Ukraine collapsed. Second, as the economy deteriorated rapidly in the early 1990s, most people began to focus their attention on key issues of daily survival and thus had little time or interest in more marginal issues, such as environmentalism. This lack of interest in the nuclear question would also have been reinforced by the sense that the problem had already been resolved with the Ukrainian moratorium resolution of 1990. Finally, as the implications of independence began to sink in, both politicians and the public began to recognize that the nuclear power stations provided them with independent control over their own energy supplies and thus enhanced both the republic's sovereignty and chances for economic survival. Thus, despite continuing concerns about the safety of Soviet-built nuclear reactors, the movement against nuclear power in Ukraine had virtually disappeared from sight by the beginning of 1992.

Anti-nuclear demonstrators join the May Day parade in Kiev, 1990.

IAEA president, Hans Blix, meets with Yurii Shcherbak, president of Zelenii svit, at a press conference in Kiev, 1990.

Zelenii svit's new offices in Kiev receive a blessing, June 1990.

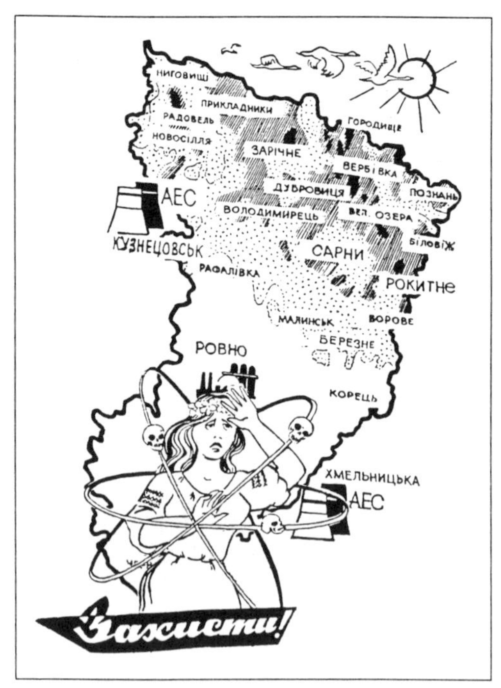

The view from Rovno Oblast—ringed by nuclear power stations (AECs in Ukrainian).

Protesters weather the cold to make their voices heard against the Khmelnitsky AES.

An environmental protest in Kazan, Tatarstan, March 1990.

"Down with the AES!" chant protesters in Kamskaya Polyana, Tatarstan, 1989.

"Let the CPSU live in the Chernobyl AES"—a typical slogan in Russia, Ukraine, and Lithuania.

CHAPTER 4

The Battle against Khmelnitsky AES: A Close-up View of Mobilization in Ukraine

Prior to 1988, the Khmelnitsky oblast was known as a quiet, sleepy region in central Ukraine. Located just to the east of the boundary of pre–World War II Ukraine, the Khmelnitsky region had been part of the Soviet Union since the early 1920s, and the population was long accustomed to life under Soviet rule. The oblast was largely rural, with more than half of the population living outside what the Soviets considered urban centers.[1] Even during the early years of perestroika, it was known as a conservative stronghold with a quiet, docile, and largely apolitical population.

In the late 1980s, however, all this changed. The introduction of glasnost on questions of nuclear power in 1988 made the population suddenly sit up and take notice of its own situation and the nuclear threat confronting the oblast. Not only was an immense nuclear power facility sprouting up in the town of Netishin at the northwest corner of the oblast, but a second nuclear power station was already in operation in the neighboring oblast of Rovno, and plans were under way to continue expanding that station as well. Thus, the residents of Khmelnitsky oblast found themselves threatened by two immense facilities located within 140 kilometers of each other. This situation, combined with the flood of information on the dangers of nuclear power and the growing knowledge of the cover-up of the magnitude of the Chernobyl accident, led the residents of this previously peaceful region to mobilize and fight Moscow's plans to expand nuclear power facilities in their neighborhood.

The fight against the Khmelnitsky nuclear power station began in earnest in late 1988. Prior to that time, opposition to the station was confined to quiet mumblings within the population. Within a year of the Chernobyl disaster, individuals in the towns surrounding the expanding Khmelnitsky AES began quietly discussing their worries about the new station with other concerned neighbors. Residents of the towns of Shepetovka, Netishin, Ostrog, and Slavuta were most concerned about this issue. In local factories of Shepetovka, several attempts were made in 1987 and early 1988 to circulate petitions opposing the Khmelnitsky AES. At that time, however, such behavior was considered antistate, and petitions were confiscated.[2] Thus, while popular opinion was beginning to shift in an anti-nuclear direction by 1987, no informal groups were established at the local level to fight the station, and no major actions were undertaken at this time.[3]

According to anti-nuclear activists I interviewed, the key factor in mobilizing society against the nuclear power station was the flood of information that suddenly deluged the population after January 1988. Because restrictions on discussing nuclear power issues in the press were significantly reduced at that time, the population was suddenly confronted with a variety of new opinions and information on this issue. People who had not even thought about the new nuclear power station in Netishin or the government's plans to expand it, suddenly learned that the station had experienced major construction problems, numerous small accidents, and unscheduled shutdowns of its single completed reactor, and was possibly defectively constructed. Furthermore, they learned that the government intended to construct an additional five reactors at that site.

In addition to learning of the government's plans for their neighborhood, the residents of Khmelnitsky also learned of the true magnitude of the Chernobyl disaster. In 1988, article after article was published exposing government lies on the extent of contamination in the republic. Residents learned that regions that the government had claimed were clean were actually unlivable[4] and that the government's claims that the accident evacuation had been well managed were false.[5] New horror stories about the levels of contamination in the soil, water, and food supply emerged every day in the Ukrainian press. Articles claiming huge increases in cancer and other disease rates in the republic were everywhere, and photographs of deformed plants and animals shocked the population. In Ukraine, the increase in glasnost in the nuclear power area led not only to an exposé of the true consequences of the Chernobyl accident, but it also triggered a wave of sensationalistic and often irresponsible journalism which terrified these long-sheltered populations.

While the new information on nuclear power and the Chernobyl accident generally emanated from the newly activated intellectual elites of Kiev and West Ukraine, it served to mobilize an entirely different population. Outside the major urban centers of Kiev and Lvov, pockets of intellectuals were few and far between. In the oblast of Khmelnitsky, it was the workers, schoolteachers, and low-level bureaucrats who were swept into this frenzy of anti-nuclear activism. The magnitude of the threat confronting them dragged them out of their long lethargy and into a determined, angry crusade against those government organs who would impose nuclear power on their oblast.

In the town of Shepetovka, one of the major industrial centers of the oblast, angry citizens finally began to organize against the Khmelnitsky AES in March 1989. According to interviews with local activists, the decision to organize against the station arose as a response to Moscow's pronuclear propaganda campaign carried out during 1988 and early 1989. Numerous interviews have revealed that teams of nuclear specialists were sent out to the factories and social organizations in the oblast to "educate" people on the desirability and reliability of nuclear power. During these educational sessions, specialists insisted that nuclear power was absolutely safe and necessary. By 1988, however, the population was being flooded with new and contradictory information, and thus these sessions angered many in the audience and reinforced growing suspicions that government officials were simply not to be trusted.

After one such session in early 1989, several local residents, including Pavel Opanasiuk, a low-level bureaucrat in the agricultural industry department, Taras Nagulko, a local doctor, and Yurii Gavriluk, a teacher, got together and decided to confront these nuclear specialists with some of the information being published in the press at the next such session. Thus, on March 15, 1989, the nuclear specialists found their plans for a quiet educational session at the Railroad Workers' Club in Shepetovka quite rudely interrupted. Rather than sitting quietly and accepting the specialists' claims, a handful of local opponents of nuclear power began asking questions and accusing the specialists of falsifying information. Soon other local residents began to sit up and interject their own questions and doubts into the discussion, and within no time, the specialists had lost control of the session. They were so shocked by the reaction that eventually they called in the Communist Party first secretary of the region, Zagalevich, to bring the meeting back under control.

At the end of this session, concerned residents stayed on and decided to elect an initiative group to organize an anti-nuclear organization in their region. They named the group "Nabat [meaning alarm bell]—The Committee to Halt Construction and Close Khmelnitsky AES," and thirteen people, including Pavel Opanasiuk, Taras Nagulko, and Yurii Gavriluk, were elected to lead the initiative group. The platform of this new group consisted of a number of points, including halting construction of the Khmelnitsky AES, carrying out a media campaign against the station, staging protests and demonstrations, and eventually bringing about the closure of the entire station.[6]

Initially, however, local authorities showed little sympathy for the anti-Khmelnitsky cause. As mentioned earlier, attempts to circulate petitions in 1987–88 were quickly halted. Similarly, the thirteen people who were elected to Nabat, the first informal association to emerge in Shepetovka, were not pleasantly received by local officials. Within days, each Nabat member was visited by officials from the USSR Ministry of Internal Affairs and warned to halt their independent, "antistate" activities immediately. About half of the initiative group, in fact, responded to these threats and ceased all activities. Only a tiny core of perhaps a half-dozen people, including the original organizers, Opanasiuk, Nagulko, and Gavriluk, remained active and continued their activities despite continued harassment and threats from the long-feared Ministry of Internal Affairs.[7]

From March 1989 on, the Shepetovka activists met informally in Opanasiuk's office at the district government (raiispolkom) building. Because they posted fliers advertising their meetings as open to the public, many curious observers attended the sessions. Communist Party and local government officials were frequent visitors during the first several months of the Nabat group's activities. Initially, the party and government officials attempted to continue their pro-nuclear propaganda campaign and tried to convince the activists of the safety and desirability of the Khmelnitsky AES. After a few months, however, they recognized that their attempts were futile and ceased to attend the open sessions.

Within weeks of forming the anti-nuclear initiative group, activists decided to declare themselves the Shepetovka chapter of the all-Ukraine organization Zelenii svit. Shepetovka was not the only town in the area, however, to boast its own informal anti-nuclear association in spring of 1989. People were waking up all over the oblast, and activists in Ostrog, Netishin, and Slavuta were beginning to organize their own local campaigns against the Khmelnitsky station. In the neighboring towns of Ostrog and Netishin, the anti-nuclear movement was mobilized almost single-handedly by an energetic and determined schoolteacher, Tatyana Matunina.[8] Having witnessed the death of her own brother as a result of a blood disease, Matunina was deeply concerned by reports of undisclosed radioactive contamination in Ukraine and the flood of articles in the local press indicating that the impact of the Chernobyl accident on the population's health was likely to be much higher and more deadly than the government had originally led people to believe.

In late 1988, Matunina and a nurse from the medical staff of Khmelnitsky AES, Zina Gomanova, managed to organize an open meeting on the impact of Chernobyl on the health of the residents of Khmelnitsky oblast. Medical personnel from the station agreed to participate in the meeting and were even able to arrange access to the station's auditorium for this session. Union representatives from most of the local workplaces were invited to attend. This was apparently the first opportunity local residents had to discuss their concerns about Chernobyl and nuclear power openly. As in Shepetovka, this session led to the creation of an initiative group that would organize an anti-nuclear association.

During the spring of 1989, a tiny core of activists attempted to publicize their concerns among the Ostrog and Netishin populations. Due to Matunina's position as a teacher, much of their effort was focused on schools, and numerous sessions were held with teachers to discuss the nuclear power issue and to convince them to join the cause. During the spring, the initiative group held a second open session in the Khmelnitsky AES auditorium to discuss the environmental situation in their oblast. While the session included discussion of a variety of local environmental problems, the real focus of the meeting was on the nuclear threat to the region. At this session, the initiative group decided to organize its first protest "meeting" and began putting together a list of meeting resolutions.[9]

When the Ostrog greens appealed to local authorities for permission to hold a public meeting, however, their appeal was initially rejected. Local authorities requested that the activists instead hold a roundtable with station administrators and staff. Matunina and other members of the initiative group agreed, and soon thereafter an open roundtable was held at the station. While attendance was high, the quality of discussion was apparently quite low. As usual, the station administrators dominated the session, and the activists felt that they were simply being bombarded by state propaganda. Opponents of the station were given little opportunity to speak their mind. When the greens attempted to contest information given out by the station director, a heated argument erupted, and the roundtable was hastily brought to a close.

In June 1989, the Ostrog greens were finally given permission to hold a mass rally.[10] By this time, they had decided to declare themselves the Ostrog chapter of Zelenii svit, and thus Zelenii svit was the official sponsor of the rally.[11] It was held June 6, 1989, in the Netishin stadium, and, according to several reports, attendance was close to 25,000.[12] Matunina and other members of the initiative group had prepared "meeting" resolutions in advance which included a call for an expert commission to review the safety of the first reactor at Khmelnitsky AES and the plans to expand the station. After the rally, Matunina sent a series of telegrams to the USSR Congress of People's Deputies demanding that they send out an expert team. No answer was ever received.

Members of the Shepetovka chapter of Zelenii svit also attended this rally, and one of their leading activists, Nagulko, presented a speech. The Shepetovka chapter of Zelenii svit also undertook its first mass rally during the spring of 1989. While a full report on the "meeting" is not available, several sources confirm that an environmental rally was held in Shepetovka on April 16, 1989. Meeting resolutions included a variety of demands on environmental issues, including the demand to halt construction of the second reactor at Khmelnitsky AES and to send out an expert commission to determine whether reactor #1 actually met international safety standards. According to the introduction to the list of meeting resolutions, the meeting was directed primarily at workers' collectives in the area, and workers were asked to organize themselves in their workplace against the Khmelnitsky AES.[13] Because no press report on this meeting has been found, however, and because most people point to the April 26 meeting in Netishin as the first real mass protest in the region, it seems likely that this meeting was small and attendance was low.

The mass meetings held during the spring of 1989 in Shepetovka and Netishin came as quite a shock to the residents of this quiet oblast. Many people recall these meetings as the beginning of the politicization of the region. People who had never before considered opposing state policy in any way began to recognize the possibility of participating in this movement and perhaps shaping state policy. Thus, the summer of 1989 was an exciting, even euphoric, period for these new activists.

Before discussing the events of the summer of 1989, however, it should be noted that while workers and other nonintellectuals were organizing in Ostrog, Netishin, and Shepetovka, a parallel mobilizational effort was under way among the journalist community in the oblast, particularly the journalists in Slavuta. While there was no informal press in Shepetovka, Ostrog, or Netishin prior to 1991, the local weekly newspaper, *Trudyvnik Polessa,* an organ of the Slavuta city party organization and city soviet, took the lead in publishing articles on nuclear power and the Khmelnitsky AES during 1989. Several journalists on the staff as well as the editor, Valerii Basyrov, were extremely sympathetic to the anti-nuclear view, and during 1989 and early 1990 their paper became the main forum for public discussion on this issue. A series of articles on the station, as well as activities of local anti-nuclear groups, were published by Basyrov. While the editor claims that both sides of the issue were presented, the available evidence indicates that the newspaper had a distinctly anti-nuclear orientation.[14]

During 1989, Basyrov and other journalists from *Trudyvnik Polessa* began to reach out to journalists in other towns of the oblast, and a network of anti-nuclear journalists began to coalesce. The journalists viewed their role in the mobilization of the anti-nuclear movement in the oblast as central. In addition, this network of journalists also undertook a collection of signatures against the Khmelnitsky station in 1989 and managed to collect approximately 20,000. The signatures were sent directly to Mikhail Gorbachev, and an open letter to Gorbachev, along with several pages of signatures, was published in *Trudyvnik Polessa.* Gorbachev responded by advising the journalists to contact the USSR Ministry of Atomic Energy.

The response of local authorities to the journalists' audacity was quite negative, and there were published calls to publicly condemn the journalists and to punish them in some way. In fact, local authorities did not welcome the journalists' input in the nuclear power question, and the more conservative papers in the area published a number of attacks on incompetent journalists who dared to speak outside their area of professional competency. Again and again, the journalists were warned to leave scientific matters to scientists and to halt their anti-nuclear crusade.

Thus, in the towns surrounding the Khmelnitsky station, two parallel and complementary movements were under way during 1989. On the one side, a network of journalists emerged, determined to alert the population to the true nature of the nuclear threat in their oblast. On the other, several cores of nonintellectual activists sprung up in the towns surrounding the station and undertook to organize and activate the local populations. Apparently, however, these two parallel networks were never linked in any way. While the articles published by Basyrov and other concerned journalists provided the activists with ammunition for their fight, the journalists did not play a direct role in organizing the anti-nuclear movement.

While chapters of Zelenii svit played the key role in mobilizing the residents of Shepetovka and Ostrog against the Khmelnitsky AES, it was not always ecological clubs that led the movement. In fact, the leadership of the movement differed from town to town. In Ostrog, Zelenii svit was the only informal group of any significance and thus led the anti-nuclear battle single-handedly. In Shepetovka, Zelenii svit was the first informal group to emerge; it was assisted, however, by members of the Rukh movement after it emerged in 1990. In contrast, in Slavuta, the journalists initially acted through their own informal network, then in early 1990 they joined the newly created local chapter of Rukh en masse. They continued their anti-nuclear crusade from within Rukh for a short while and then shifted to other issues. Finally, in Netishin, the authorities remained extremely conservative, and even Zelenii svit was viewed as too suspect to be permissible during much of 1989. Thus, during the early mobilizational period, the more politically acceptable Shevchenko Language Society was the only group of significance in Netishin, and it led the local fight against Khmelnitsky station.

In fact, during 1989, the names of the informal groups didn't matter much. They were all involved in the same crusade, and their form often depended merely on what local authorities were willing to put up with and what the handful of activist citizens in that particular town wanted to call themselves. It should be stressed that while the protests brought thousands of people out onto the streets, membership in informal organizations remained very low, and it was really only a few energetic individuals that kept the battle going in each town. In 1989, however, the struggle against nuclear power and attempts to promote Ukrainian language and culture were the only significant activities of the newly formed informal organizations in the Khmelnitsky oblast. Given the political climate of Ukraine in 1989, more overtly political demands were considered far too dangerous to risk, and most of the newly activated population of Khmelnitsky oblast focused their attentions on the relatively safe and apparently apolitical nuclear power issue.

Throughout 1989, the anti-Khmelnitsky movement grew, eventually peaking during the summer of 1990. During the summer and fall of 1989, anti-nuclear activists in the region circulated dozens of petitions among the local residents and collected tens of thousands of signatures. It should be noted, however, that the efforts of the various activists were very rarely coordinated, and thus the wording of each petition depended on its author and the demands vary somewhat. In 1989, these petitions were always sent directly to the president of the USSR, Mikhail Gorbachev, the USSR Council of Ministers, and the USSR Congress of People's Deputies, with copies occasionally being sent to the Ukrainian Council of Ministers and the regional party organization.

While local party authorities were not terribly sympathetic to anti-nuclear demands during 1987 and 1988, by 1989 a shift in the attitude of some authorities began to be discernible, and by early 1990 many local party organizations had begun to openly support the anti-nuclear cause. The reason for this shift in party attitudes was twofold. On the one hand, local party officials were also residents of the neighborhood, and just like their neighbors, they were exposed to the flood of new information on the dangers of nuclear power and the damage caused by Chernobyl which was published after January 1988. Thus, many of these officials were reportedly seized by fear for their own families and homes.

On the other hand, the local party organizations began to recognize the power of the anti-nuclear issue in mobilizing the population. By the summer of 1989, numerous anti-nuclear rallies were being held in the towns surrounding the station, and the previously docile population of the oblast was turning out in droves to support the anti-Khmelnitsky cause. In addition, the petitions and the publication of an open letter to Mikhail Gorbachev, which appeared in *Trudyvnik Polessa* in 1989 along with claims of over 20,000 signatures, is reported to have awakened party officials to the level of popular discontent in the oblast.[15] Thus, during 1989 and 1990, sympathy and support for the anti-Khmelnitsky cause grew substantially among local officials.

During the spring of 1990, the strategy of the anti-nuclear movement underwent a tremendous change. Up until this time, local activists had focused their efforts on demonstrating their opposition to the nuclear power station to decision makers in Moscow. Their tactics were limited to mass demonstrations and petition drives, and the stacks of signatures they collected were always sent to Moscow in an effort to alter state policy. All this changed in 1990 with the introduction of multicandidate elections at the republic and local levels.

During January and February of 1990, candidates were chosen for local, city, and oblast soviets, as well as for the Ukrainian Supreme Soviet, and electoral campaigns got under way. Unfortunately, the budding informal groups in Ukraine were still quite weak during the winter of 1989–90, and, unlike in Lithuania, they were unable to force Ukrainian government officials to provide them with equal access to the electoral process. Thus, whereas the Lithuanians were able to nominate candidates from informal groups such as Sajudis to both the all-union and local elections in 1989 and 1990, respectively, the old regime in Ukraine remained strong enough to deny this privilege to the new informal groups of the republic. Thus, while local chapters of Zelenii svit, Rukh, and the Shevchenko Language Society attempted to nominate their own candidates in the local elections of 1990, the electoral commissions at all levels refused to register these independent candidates. Only candidates from officially acceptable social organizations were permitted to register.

While a number of Rukh candidates were able to make it onto the ballot by quickly finding an alternative official organization willing to nominate them, the members of Zelenii svit were not quite as successful. Rather than pragmatically finding an alternative route to nomination, the all-Ukraine Zelenii svit association chose to protest Kiev's unfair policies and to attempt to force the government to back down. Unfortunately for them, their protests were not heeded and Zelenii svit candidates were left out in the cold. While some managed to be nominated by other organizations, members of the Zelenii svit secretariat in Kiev admitted later that their handling of the 1990 elections represented a major strategic error.

In Khmelnitsky, the only member of the anti-nuclear crusade to be nominated for membership in the Ukrainian Supreme Soviet was Taras Nagulko, one of the original founders of the Shepetovka chapter of Zelenii svit. A handful of local activists also made it onto the ballot for elections to the oblast and local soviets. The lack of genuine greens on the ballot, however, did not imply that the electoral process was irrelevant to the green movement. In fact, the elections of 1990 had an immense impact on the strategy and level of success of the anti-Khmelnitsky movement.

By early 1990, no one in the region could fail to be aware of the popular discontent with the ever-growing Khmelnitsky nuclear power station. Thus, all of the candidates recognized the political value in stating their adamant opposition to the Khmelnitsky station. Suddenly, everyone was anti-nuclear; party conservatives and progressive members of the emerging political opposition alike avowed their heartfelt opposition to the overwhelming nuclear threat in their region. With the need for local support so obvious, only a fool would have stood up and defended plans for the Khmelnitsky AES, and there is much evidence to suggest that almost no one was quite so politically naive. Thus, the candidates attempted to outdo each other in expressing their rage over the continued construction of Khmelnitsky AES and Moscow's insensitivity to local concerns.

As a result of the republic and local elections, the councils were packed with deputies who had all claimed to be more anti-nuclear than their opponents during their electoral campaigns. At the republic level, this resulted in a decision to freeze all nuclear construction in Ukraine.[16] At the local level, the elections led to a slew of decisions to take matters into local hands and halt projects deemed unacceptable to the electorate.

When the new Khmelnitsky oblast soviet met for the first time in April 1990, the members were unexpectedly confronted by a proposal to cancel the expansion of the station immediately. A vociferous opponent of the Khmelnitsky AES from Shepetovka, L. Dorofeyevna, stood up and angrily demanded that the new deputies fulfill their promise to the electorate and halt the expansion project immediately. Other deputies joined in the call for an immediate decision, suggesting furthermore that the tally on the vote be open and published in the official press. Despite the fact that no detailed proposal for halting construction of the station had been prepared in advance and thus none of the ramifications of such a decision considered, the deputies felt under considerable pressure to demonstrate the sincerity of their campaign promises. Thus, sometime after midnight, a proposal to halt construction at the station was voted on and passed unanimously.[17]

The decision of the Khmelnitsky oblast soviet set a precedent for the entire republic, and represented a dramatic change in the strategy of greens across the republic. While the greens had focused their attentions almost exclusively on Moscow decision makers prior to the 1990 elections, after these elections attention suddenly turned to republic and local decision makers. Suddenly people demanded the right to make their own decisions on such critical questions as whether they wanted a nuclear power station in their area. Thus, while the Khmelnitsky decision was the first of its kind in Ukraine, it set a precedent that was replicated across the republic.

While the Khmelnitsky oblast soviet made the decision to cancel the expansion of Khmelnitsky AES, no one was quite sure at the time what this decision meant. Because Moscow's authority in the republics was tumbling, local decision makers saw an opportunity to increase their own decision-making power. They were unsure, however, how their actions would be received in Moscow and whether their decisions would have any influence at all.

In fact, the first sign of an official reaction to the Khmelnitsky oblast soviet decision came from the nuclear power station itself. Immediately following the announcement of the oblast decision to halt construction at Khmelnitsky AES, top members of the station's staff arrived in the oblast capital to try to convince deputies of the error of their ways. Members of the two oblast soviet commissions most directly concerned with the station, the commissions on ecology and industry, were immediately whisked off to the station for a tour and roundtable meeting with workers and administrators of the station. The deputy chair of the commission on ecology, Yurii Reznikov, a strong opponent of the AES, spoke to workers during his visit and tried to convince them of the health threat of an atomic power station. While he found some workers sympathetic to his concerns, the majority clearly opposed the oblast soviet's decision to curtail work at the Khmelnitsky station.

Soon after the roundtable at the Khmelnitsky AES, Reznikov and several other members of the commission on ecology traveled to Moscow to present the oblast decision to Prime Minister Nikolai Ryzhkov, and chairman of the Fuel-Power Complex, Lev Ryabev (as dictated in the April resolution of the Khmelnitsky oblast soviet). The delegation failed to convince Moscow decision makers of the validity of their decision, however, and returned empty-handed.

Shortly thereafter, on May 17, 1990, a high-level delegation from Moscow arrived in the oblast to meet with members of the ecology commission and convince them to reconsider the decision to cancel the expansion of Khmelnitsky AES. The delegation included the deputy chair of the USSR Gosatomnadzor (Atomic Control Commission), the USSR deputy minister of power, and representatives of top design and construction institutes in Moscow. Although they had planned to meet quietly with members of the oblast ecology commission in a small room at the oblast ispolkom (executive-committee) building, local residents got word of their impending visit, and when the delegation arrived it was confronted by a huge crowd gathered in front of the building demanding that the session be open to observers. Several opponents of the station, who feared that the Khmelnitsky oblast soviet was about to back down to Moscow's pressure, had even begun a fast in the square the day before. Local citizens were determined to prevent the Moscow delegation from unduly influencing the ecology commission of the oblast soviet; they eventually succeeded in having the session moved to the large auditorium at the local symphony hall. The session proved to be a lively one, with local residents arguing vehemently against construction at Khmelnitsky AES and the Moscow delegation futilely trying to convince them that continued construction was in their own best interest. The end result, however, was a declaration by the delegation that they had no authority to confirm a decision taken at the oblast level and, in fact, that they believed the oblast decision to be irrelevant and invalid. Throughout the spring and summer of 1990, construction of the second reactor at Khmelnitsky AES continued unabated.

By June 1990, however, anti-nuclear activists were beginning to lose tolerance for Moscow's refusal to recognize the validity of the local decision to halt construction at Khmelnitsky AES. The hunger strike that started in May at the oblast capital was still under way,[18] and activists in the towns surrounding the station were becoming fed up with Moscow's lack of interest in their opinions. A series of warning strikes were carried out in factories near the station,[19] and more serious steps were threatened.

The cement factory in Zdorlbunovo threatened particularly radical measures, warning that they would halt deliveries of cement to the station and block train deliveries of other materials if construction continued.[20]

Finally, in June and July, the local chapters of Zelenii svit and Rukh jointly called on residents to join them in picketing the Khmelnitsky station. The picketing organizers threatened to continue to hold mass protests around the station until construction was halted.[21] This action greatly exacerbated tensions between the anti-nuclear activists and the workers and staff at the station. While Matunina and other protest organizers attempted to prevent participants from blocking access to the station and thus endangering the operation of the station, the protestors occasionally got out of control, and several attempts were made to completely blockade the station. In the end, the protestors agreed to a compromise and allowed workers to go freely to and from the operating first reactor but obstructed their passage to the construction sites of new reactors.[22]

The director of the station, Victor Sapronov, reacted extremely negatively to the picketing of the station and sent telegrams to Moscow, Kiev, and the oblast capital demanding that action be taken to remove the picketers. Official reaction to the picketing in Kiev and the Khmelnitsky capital was also quite negative. Nagulko, Ukrainian Supreme Soviet deputy from Shepetovka, and Dorofeyevna, Khmelnitsky oblast soviet deputy from Shepetovka, both received official reprimands for participating in the protests. Matunina, Opanasiuk, and four other members of Zelenii svit and Rukh who had helped organize the action were also called to a court hearing and fined for crossing the "sanitary zone" of the nuclear power station.

In late June, the Khmelnitsky oblast soviet attempted to clarify its hasty April decision by passing a resolution "on implementing the decision of April 6, 1990." This resolution acknowledged the extensive consultation with members of the ecology commission as well as the Khmelnitsky AES administration, and attempted to avoid the huge losses in investment that would be caused by completely halting construction at the station. Thus, this resolution allowed construction to continue on reactors #3 and #4 just to the point where safe conservation of the structures could be achieved. The resolution also acknowledged Sapronov's concerns about the dangers of picketing at the station and forbade deputies from participating. A third resolution in September 1990 went even further and forbade picketing and mass protests within a three-kilometer radius of the station.

In August 1990, however, the entire anti-Khmelnitsky movement began to simmer down and fade away. The Ukrainian Supreme Soviet's moratorium on the expansion of nuclear power in Ukraine convinced most people that the battle had been won and that there was no need to continue the fight. Moscow had lost and Ukraine had won. Despite the fact that construction on the second reactor at Khmelnitsky AES continued and that the Supreme Soviet moratorium resolution was worded so vaguely that no one seemed to know how it applied to stations already under construction, local residents rapidly lost interest in the issue after August 1990.

By early 1991, interest in the Khmelnitsky AES question had dropped so low that members of the oblast soviet felt safe in radically modifying their initial April 1990 resolution. In a fourth resolution on this question, issued January 9, 1991, the oblast soviet announced that the second reactor at the station would be completed in order to save the government's investment in the station and then would be "conserved" pending a later decision by an expert review team.[23] Thus, the story came to a close. Construction of the second reactor was in fact completed soon thereafter. The expert review team ordered by the oblast soviet never materialized due to lack of funds and lack of interest in the question by members of the oblast soviet.

While a handful of key activists, such as Matunina and Opanasiuk, continued to pursue their campaign against the Khmelnitsky station, they found the local population unreceptive after August 1990. After a summer of intense mobilization, people were tired of taking to the streets. Furthermore, for many the battle had already been won. Ukraine had demonstrated its authority to dictate its own policy, and Moscow's control over the nuclear power issue had apparently been lost forever. Whether or not the station actually disappeared, many people felt that their demand for the rights of self-determination had been achieved and that the struggle need go no further.

Interestingly enough, after the Ukrainian parliament overturned its moratorium on nuclear construction in December 1993, the Khmelnitsky AES was one of the first stations targeted for renewed construction and expansion. A presidential decree dated February 23, 1994, called for the second reactor to be brought into operation in 1995, followed by a third in 1998.[24] This decree was later confirmed by a government resolution on May 12, 1994.[25] At present, construction is again under way at the Khmelnitsky station, and while a handful of local citizens continue to oppose the expansion and operation of the station, there is no sign of mass opposition. As elsewhere, the nuclear power issue is no longer a potent mobilizational factor in the region.

Conclusions

The anti-nuclear power movement in Khmelnitsky was significant for its role in activating a long acquiescent and docile society. Prior to the anti-nuclear protests of 1989, there was no sign of any popular opposition to state policy on any issue. Thus, mass participation in anti-nuclear rallies and petition drives represented a major change in how the population viewed its relationship to the state.

It is important to note, however, that in the countryside, participation in the anti-nuclear movement was generally premised on genuine fears about the safety of neighboring nuclear power stations and the threat they posed to home and family. Thus, mobilization against the station did not translate directly into political mobilization. Even in 1990 and 1991, democratic organizations remained extremely weak in the Khmelnitsky oblast. The lack of influence of the fledgling democratic movement in the region can be discerned in the conservative composition of both the oblast and local soviets that resulted from the 1990 elections.

Additionally, Ukrainian national identity proved unexpectedly weak, and while the national question was linked to anti-nuclear demands, nationalism never eclipsed the nuclear power issue. Thus, while activists demanded that Ukraine be protected from Moscow's dangerous plans to expand nuclear power in the region, popular fears about nuclear power were quite real, and the nuclear power issue was not simply a surrogate for nationalist aspirations.

CHAPTER 5

Russia: The Demand for Local Self-Determination

Introduction

While the Lithuanian and Ukrainian anti-nuclear movements were intimately intertwined with aspirations for national sovereignty and independence from the Soviet Union, the movements that emerged in Russia were of a quite different character. Rather than acting as surrogates for forbidden Russian nationalist demands or acting in tandem with a strong nationalist movement, the Russian anti-nuclear movements had little or no connection with the nationalist cause.

Like Ukraine, Russia possessed not one nuclear power station but rather a multitude of stations scattered across the entire republic. Because Soviet planners had decided in the mid-1970s to concentrate nuclear power stations in the European USSR (due to the density of industry in this region), the majority of Russia's nuclear power stations were located in the western portion of the republic. Thus, it would not have been surprising if the Russians had duplicated the experience of their Ukrainian neighbors and created a powerful all-republic umbrella organization to fight the expansion of nuclear power in Russia. This, however, did not occur. Rather than uniting in their anti-nuclear cause, the Russian movements remained almost totally isolated from one another and an all-republic organization failed to emerge.

In fact, the movements that emerged in Russia would be most accurately characterized as NIMBY movements—that is, as the protests of local residents vocally demanding that dangerous industrial facilities be built "Not In My Back Yard!" From 1987 to 1990, anti-nuclear protest groups emerged around virtually every nuclear power station in operation, under construction, or planned in the Russian republic. Furthermore, similar protest movements emerged in the republic around a variety of industrial facilities thought to be dangerous to the health of local residents.[1] While some attempts were made to create networks to link these numerous NIMBY groups, during the perestroika period none succeeded in forming a powerful umbrella to unify the environmental and anti-nuclear forces of Russia.[2]

The anti-nuclear and other NIMBY movements of Russia were among the first movements in the republic to emerge and attract popular support. In many regions, protests against nuclear facilities and other dangerous industrial plants were the first to bring people out onto the streets and to mobilize long-dormant sectors of society. They were the beginning of the politicization of society and were built on the demand for greater local decision-making rights. These anti-nuclear campaigns alerted local populations to their powerlessness in the face of Moscow's monopoly on decision making, and began to create popular support for increasing local decision-making rights vis-à-vis the central government.

It is in Russia that the competition between different levels of authority became most obvious and pervasive in the anti-nuclear case. While initial appeals of anti-nuclear activists focused attention on USSR organs and individuals, in 1990 attention shifted to the city, oblast, and republic level. Following the 1990 elections, local soviets were quick to decree the cancellation of nuclear projects in their region. In some cases, the local councils claimed that this was their right according to the new USSR law on local self-determination. In other cases, the local soviets appealed to the Russian government to confirm their decisions. In the cases where the Russian government did step in to confirm local cancellation decisions, however, they often found themselves in direct conflict with USSR organs refusing to acknowledge the validity of the local decisions. Thus, in Russia, the struggle between competing levels of authority was openly played out.

In this chapter, I will present a detailed case study of the anti-nuclear campaign waged against the Gorky nuclear facility. This study not only illuminates the dynamics of social mobilization in Russia but also provides an excellent example of the competition between levels of authority that occurred in the nuclear power sector in 1990–91. I conclude the chapter with some thoughts on the linkage between anti-nuclear power platforms and nationalism in Lithuania and Ukraine and why that linkage was not duplicated in Russia.

The Battle against Gorky AST

Gorky, now Nizhny Novgorod, is probably best known as the city in which Andrei Sakharov was forced to live in internal exile for his outspokenness on the question of human rights and democracy in the USSR. It is a bleak city, whose tiny ancient center (complete with a magnificent walled fortress) is completely overwhelmed by the haphazard industrial development of the region. What was once a beautiful town is now one of the most heavily industrialized and polluted cities in the former USSR. Smokestacks are visible in all directions, and soot covers everything. The air reeks of foul smelling chemicals being spewed out of the hundreds of smokestacks. Environmental protection was clearly not a top priority in the development of Gorky.

The city was long closed to foreigners, due to the high concentration of military-industrial production in the region.[3] As a center dominated by military concerns, it was known as a conservative stronghold during the Brezhnev era. It is thus not surprising that local residents were reticent to mobilize on political platforms during the early perestroika period. Given the dramatic environmental degradation of the region, however, the construction of yet another dangerous industrial facility in the Gorky area offered an obvious target for quick mobilization of local residents.

The decision to build a nuclear power facility in the Gorky oblast was adopted in 1979. The Gorky nuclear facility, however, was not to be a typical AES but rather a brand-new type of reactor designed specifically to provide heat to the region. This new atomic heating station, known by its Russian initials AST, was jointly designed by the Gorky Experimental Machine Building Design Bureau (known as OKBM) and the Leningrad Design Institute of Power Technology. The final design for this innovative new type of nuclear reactor was put together by Academician Fedor Mikhailovich Mitenkov of Gorky OKBM.

During the years prior to the Chernobyl disaster, the Soviets were quite proud of their new design for a nuclear heating facility. Articles mentioning the Gorky project praised it as a major step forward in nuclear technology.[4] The Gorky station was to be the first of its kind in the USSR, but other ASTs were also planned for Voronezh and Archangelsk.[5] Because heat cannot travel far without dissipating, ASTs could only be built in locations where it was possible to build in close proximity to areas of high population concentration. Thus, the Gorky AST was sited a mere five kilometers outside the boundaries of the city.

During the early 1980s, the president of the USSR Academy of Sciences, A. Alexandrov, became quite excited about the idea of using nuclear power facilities to provide heat. Another new design, known by its Russian initials ATETS, was developed which would provide both heat and electricity to areas of high-population density. Alexandrov viewed nuclear power as the answer to the heating shortage in the USSR and proposed that these ATETS and ASTs be built on the outskirts of literally hundreds of Soviet cities.[6] In 1983, the Soviets announced to the world a major redirection of their nuclear power program—one that vastly expanded its use in the production of heat.[7] The Twelfth Five-Year Plan (1986–90) included plans to build a number of such nuclear heating facilities in Russia, Ukraine, and Belarussia.[8]

The Gorky AST was viewed as the first test of the new AST design. Construction began in the early 1980s, and the two Gorky reactors were expected to go into operation by 1990.[9] The Soviet planners, however, did not take into account the possibility that the docile Gorky population might rise up someday and prevent this station from ever producing a single unit of heat.

Gorky Awakens

As perestroika slowly began to make its way into remote cities and towns outside the more cosmopolitan centers of Moscow and Leningrad, tentative attempts to speak more openly and to form informal associations began to become evident in the oblast of Gorky. As in many other regions, a local chapter of the Komsomol youth organization took the first step toward testing the possibilities of perestroika. At a Komsomol meeting at the Scientific Research Radiophysics Institute (NIRFI) on October 28, 1987, a decision was made to publish a journal which would focus on political topics and issues of concern to Gorky residents. The newsletter was entitled *Point of View* (*Tochka zreniya*) and began to appear in November 1987. The editorial board was particularly concerned with uncovering the Stalinist legacy and questioning the safety of the ever-growing nuclear power facility in Gorky. While published somewhat irregularly, this newsletter represented a dramatic attempt to test the boundaries of glasnost.

In Gorky, it was not until May 1988 that the first informal club was formed. This club, known as Avangard, was initiated by Stanislav Dmitrievsky Klokov. It defined itself as a political club and set as its first tasks (1) removing Zhdanov's name from streets and buildings in Gorky and (2) fighting the construction of the nuclear heating station on the outskirts of the city. The club was limited to a small group of intellectuals but nonetheless was viewed as a real threat by local authorities. When the club tried to stage a small rally in May, Klokov was arrested, fined, and deprived of his CPSU-party card. (The card was later returned, however.)

After Avangard announced its existence in May 1988, a number of small clubs were quick to surface. In May alone, a local branch of Memorial opened, along with a cultural-historical group calling itself the Council on the Ecology of Culture, an organizational committee for the formation of a People's Front, and a branch of the Democratic Union. All of these groups were very small, but their formation and their willingness to publicize their existence represented a dramatic step forward for Gorky.

Throughout 1988 and early 1989, these groups were consistently harassed by local authorities. Steps were taken to prevent meetings, rallies, and petition drives. In 1989, the four democratic groups mentioned above were even denied permission to form a column in the May Day demonstration. The column was attempted nonetheless and resulted in the arrest of the group leaders. They were later taken to court and fined for their insubordination.[10]

In June 1988, an attempt was made to bring these new activists together in order to forge contacts and to create the basis for future cooperative actions. The meeting was held in a dormitory at Gorky Polytechnic Institute, and approximately forty people attended. Ten groups were said to be represented at the meeting, although it is important to note that many of these groups were so small that it would be misleading to think of them as groups. Members of the editorial board of the progressive Komsomol newsletter *Point of View* both participated in the session and covered it for their newsletter.[11] Their report notes that some of the groups consisted of only one to two members, and some even had no permanent membership. It also seems apparent from this report that the meeting was not a total success. There were numerous disagreements on the direction in which the democratic movement in Gorky should develop, and attempts to compose a single platform for the movement were unsuccessful. A second meeting was held on June 22, but little more was achieved.

While the democratic movement was slowly emerging, the roots of a much more powerful movement began to grow. Again, the Komsomol chapter at the physics institute, NIRFI, provided the impetus for this new movement. Ever since the founding of the independent newsletter, *Point of View,* members of the NIRFI chapter of the Komsomol had been active in promoting their anti-nuclear views. Opposition to the Gorky AST gradually spread throughout the institute, and on March 15, 1988, a vote was taken at a union meeting at NIRFI on the question of whether the Gorky AST should be completed. A majority of members of the union voted against completion of the station.

In mid-June, the NIRFI Komsomol asked the Gorky *raiïspolkom* (district executive committee) for permission to hold a public rally on the Gorky AST issue in late June. Authorities wavered but finally agreed to permit a public rally on July 10. While the rally was small—only 1,500–2,000 participants—it represented a dramatic step in the activation of Gorky society. Participants marched from the October theater to the center of town, known as Theater Square. Approximately thirty speakers were heard. It is important to note that at this initial rally the focus was on providing information about the Gorky AST, and speeches were presented for both sides of the issue. City authorities, representatives of the station, as well as the design institute, OKBM, spoke in defense of GAST (Gorky AST).

On the opposition side, it was again a group of scientists from the physics institute, NIRFI, which led the campaign against the station. Three physicists, Boris E. Nemtsov, V. A. Alimov, and V. S. Troitsky, used their scientific expertise to convince the crowd that the absolute safety of the Gorky AST could never be achieved. Reports indicate that the overall tone of the rally was largely anti-nuclear, and the proponents of the station had difficulty making themselves heard.[12]

Following the July 10 rally, this small core of scientists from NIRFI and other scientific institutes began to wage a publicity campaign against GAST.[13] During the fall of 1988 and most of 1989, Troitsky and other anti-nuclear scientists published numerous articles in the local press condemning the Gorky AST.[14] While not all newspapers in the area agreed to publish their articles, the two Komsomol papers, *Point of View* and *Leninskaya smena,* were quite receptive to the anti-nuclear viewpoint. While initially, all television coverage of the Gorky AST issue was restricted to views supporting the station, on August 1 a dialogue between supporters and opponents of the station was televised locally. The television debate also asked viewers to call in with their opinions, and the result was 27 callers for the Gorky AST and 153 against.

The Populace Is Aroused

During the spring and summer of 1988, scientific opposition to the Gorky nuclear facility grew. While the radiophysics institute, NIRFI, was the real hotbed of opposition, scientists from a number of other institutes supported the anti-GAST platform as well. Opposition to the station, however, was not organized into a formal network or association. Each scientist pursued the cause in his or her own individual way. While Troitsky was writing furiously in an attempt to publicize his concerns about the station, Nemtsov began a petition drive to collect signatures opposing the completion of the Gorky AST.

Meanwhile, large sectors of society outside the specialist community were beginning to become aware of the nuclear power issue and to mobilize. One individual played a central role in mobilizing the masses against the Gorky AST: Yurii Mikhailovich Likhachev, a retired industrial engineer, began to take an interest in the GAST question almost immediately following the Chernobyl disaster. His fervent belief in the AST threat and his commitment to preventing the opening of this new station, were critical in the mobilization of a mass movement opposing the GAST.

Immediately following the Chernobyl disaster, Likhachev began to read everything he could lay his hands on concerning nuclear power, and quickly came to the conclusion that this technology was far too dangerous for the lax work ethic and conditions of the USSR. By 1987, he was writing letters to the country's leaders—Gorbachev, Ryzhkov, and even Chebrikov of the KGB—demanding that construction of the Gorky AST be halted. In response to his letters, a group of atomic specialists from the KGB visited him in his home and attempted to convince him of the inherent safety of the Gorky station. Likhachev, however, remained unconvinced and distrustful of *atomshchiki* (a commonly used term for the nuclear establishment) in general. Soon after, Likhachev was invited to a roundtable on nuclear power sponsored by the USSR Ministry of Atomic Energy and Industry. The USSR Committee on the Utilization of Atomic Power also invited Likhachev to visit them and discuss his concerns with their atomic specialists. Likhachev, however, did not believe that either of these official organs would take his concerns seriously and refused to attend both meetings.

It was at the July 10 rally that Likhachev first became acquainted with physicist Nemtsov and the scientific contingent opposing the Gorky AST. According to numerous reports, this rally represented the first step in the activation and politicization of Gorky society.[15] It is also significant as the beginning of an alliance between academics and general society, symbolized by the new connection between Nemtsov and Likhachev. From this point onward, opposition to the Gorky AST began to take on the characteristics of a mass movement. While Nemtsov had begun a limited petition drive in May 1988, the collection of signatures opposing the station was undertaken on a much larger scale after the July 10 rally. Likhachev and a number of other concerned citizens set up a stand for signature collection in the center of town, near the old fortress which now houses the local government, every Saturday and Sunday. In addition to collecting signatures, the new activists also posted placards with antinuclear slogans and information about the dangers of nuclear power, on a wall just behind the petition table. While this petition drive proceeded in a very haphazard way, it provided an excellent opportunity for the opponents of the Gorky AST to become acquainted with one another and to begin to put together an organization to fight Moscow's plan to open the station by 1990.

Unlike in Khmelnitsky, Ukraine, where the administration of the local nuclear power station took an immediate dislike to the protestors and angered them even further by snubbing them and ridiculing their concerns, the administration of the Gorky AST made every attempt to deal reasonably with the opposition. During the summer and fall of 1988, representatives from the nuclear power station and the OKBM design institute also gathered in the center of town every Saturday and Sunday to answer questions about the station and to attempt to convince its opponents of its inherent safety. The station printed numerous posters and information bulletins and set up a stand next to the anti-nuclear group. In interviews with these specialists, however, it became clear to me that this was an extremely demoralizing experience for these atomic specialists, and they felt that most people were not prepared to listen to what they had to say. They were viewed as the enemy, the emissary from Moscow, and their attempts to hold rational discussions on the safety issue with local residents proved entirely fruitless.

The local authorities were alarmed by Gorky's first mass rally on July 10, and banned all further mass meetings in the center of town. Henceforth, mass rallies were to be held in a park located outside the center of town. When Likhachev attempted to hold an unsanctioned rally in the town center on October 7, he was arrested, taken to court, and fined. Throughout 1988 and early 1989, local authorities harassed both democratic and anti-nuclear activists, and attempted to impede the formation of independent groups and the mobilization of a mass movement.

Meanwhile, a third contingent was beginning to mobilize on the anti-nuclear issue. This group was composed of women who viewed the nuclear power station as a threat to their families and children. Thus, a small group of women turned to the Avangard club in September 1988 and asked for assistance in organizing a rally under the slogan, "Women and Mothers: We Will Not Permit the Gorky AST to Open!" The rally was officially sanctioned and held in the park designated by local authorities. It was at this rally that a new informal group was created: Women against GAST. While this group was extremely small—less than a dozen permanent members—it represented one of the first attempts to organize a specifically anti-nuclear informal group.

During the summer and fall, petition drives were undertaken not only in the town center but also in factories and workplaces. Opposition to the station was found to be quite strong among factory workers, and a number of factory unions both signed petitions and sent letters of protest to Gorbachev. By August 1988, tens of thousands of signatures opposing the station had been collected through the various petition drives.[16]

Despite this clear evidence of popular antipathy for the Gorky AST, construction continued unabated, and anti-nuclear activists began to recognize the need to organize themselves and expand their tactics beyond petition drives. Thus, during the fall of 1988, Likhachev, Nemtsov, and approximately twenty other activists began to meet in various apartments to discuss organizing an informal association to push for the cancellation of the Gorky AST. By December, they were ready to hold the founding meeting of their new organization. The party *obkom*, however, refused to give the group space for its initial meeting.[17] Finally, the director of school #40 agreed to allow the group to use the school's auditorium for its founding session. While the local authorities called on the director to reverse his decision, he refused to back down and the meeting was held at the school as planned.

The founding session was held on December 25, 1988 and was attended by 154 people. The session was reportedly well-organized and well-advertised; fliers printed by the liberal paper *Leninskaya smena* were posted around town, and a radio station advertised the meeting free of charge. They decided to name the group For Atomic Safety (ZAB), and a number of previously existing informal groups quickly affiliated themselves with the new anti-nuclear organization. Thus, Women against GAST, Avangard, the Council on the Ecology of Culture, and a variety of other clubs united under the new umbrella of ZAB.

In a clear attempt to boost the authority of the new organization, the prominent scientist Troitsky was elected president. Nemtsov was a member of the central council. While attempts were made to present ZAB as a specialist and intellectual organization, its composition was actually extremely mixed. The council of twenty-eight was dominated by industrial engineers rather than scientists, and a number of leading anti-nuclear scientists, including Devyatikh and Gaponov-Grekov, remained outside the organization.

It is interesting to contrast the early attempts to unite informal activists under the democratic flag in June 1988, and the unification of activists under the anti-nuclear slogan in December 1988. While activists found it extremely difficult to agree on democratic, political objectives and to mobilize large sectors of society behind democratic slogans, the anti-nuclear cause presented a much more direct and appealing platform for mobilizing a long-dormant society. There was little disagreement on the goals of the movement: simply, to prevent the Gorky AST from opening. In addition, the fact that authorities had long-tolerated anti-nuclear activists congregating in the center of town to collect signatures and discuss the issue proved to even the most timid that participation in this movement was unlikely to lead to severe retribution by authorities. Thus, the anti-nuclear umbrella succeeded in unifying a variety of informal groups and in bringing about the activation and politicization of Gorky society.

The Reaction from Above

As already noted, local officials reacted to the emergence of informal groups and public activities with fear and hostility. After the first mass rally, which involved less than 2,000 participants, local authorities acted quickly to ban all mass gatherings in the center of town. In addition, any attempts to defy the ban were dealt with quite aggressively; organizers of unsanctioned rallies were consistently arrested, taken to court, and fined for insubordination. During 1988, continuing official hostility to independent groups and activities ensured that participation in such groups was limited to a tiny sector of society. The most daring act that broader sectors of society attempted was the signing of petitions against the construction of the Gorky AST. Because local officials initially viewed this particular activity as nonthreatening and apolitical, participation in petition drives and letter-writing campaigns against GAST represented the first and only step in the activation of the bulk of Gorky's population during 1988.

By the end of the summer of 1988, however, the petition drives were beginning to show results, with tens of thousands of signatures already collected. Local authorities began to recognize the powerful resonance of this issue among the population, and on August 22, the first official step was taken to address the concerns of local residents. At this time, the Gorky oblast soviet executive committee issued decree number 312, "On measures regarding information and control of the ecological situation in the oblast, quality of design decisions and of construction and operational work at the AST."[18] The decree consisted of two components. First, it called upon the head of the oblast Sanitary Epidemiological Station, Yepishin, and the head of the Higher Volga Regional Hydrometeorology Committee, Ryazanov, to put together a plan for systematically informing local residents of the ecological situation in their region. Second, it established a "Societal commission for control over the quality of design decisions, construction, and operational work at the GAST." Twenty scientists were assigned to this commission, including known anti-GAST scientists Devyatikh, Gaponov-Grekhov, Nemtsov, and Troitsky. The majority of members, however, were drawn from the nuclear power sector, including leading scientists and engineers from the Gorky AST and the Gorky OKBM design institute.[19]

Meanwhile, the local units of the Communist Party were apparently in complete disarray with regard to the Gorky AST question. The most powerful party organization in the region was the oblast-level chapter, known as the *obkom*. The obkom was composed, however, of 180 to 200 members and reports indicate that there was little consensus among obkom members on the GAST question. Because of this lack of consensus, the obkom attempted to present itself as neutral on this issue, and a vote on the GAST question was never taken by the group.

Nonetheless, by the summer of 1988, there was a general feeling within the oblast party organization that the anti-GAST movement was getting out of control. The success of the petition drives brought home to the party officials the tremendous popular appeal of the anti-GAST platform, and the obkom decided that steps needed to be taken to ensure that the movement progress in a safe and nonthreatening direction. Thus, during the summer of 1988, the obkom appointed Mikhail Alexandrovich Vinogradov, vice chair of the division of economic development of the obkom, to work with the environmental movement in Gorky. Vinogradov was selected for this role because of his training as a chemist and his knowledge of environmental issues. According to Vinogradov, he and other members of the obkom viewed the emerging anti-nuclear and environmental movements as alarmist, and their hope was that through reasoned debate and discussion some of the alarmist fears of these movements could be put to rest. Thus, Vinogradov's initial goal was to defuse the growing movement.

During 1988 and 1989, Vinogradov met with anti-nuclear activists and environmentalists frequently. While his initial goal was to dampen their alarm about the safety of the new Gorky AST, Vinogradov quickly recognized the futility of this approach. Noting the tremendous popularity of the anti-GAST platform, Vinogradov gave in and began supporting the anti-GAST movement. While not accepting the safety concerns of the movement, Vinogradov explains his change of heart in terms of the need to heed public opinion on key issues such as nuclear power and the possibility that a traditional heating station in Gorky might prove more economical. Likhachev and other members of the nonspecialist component of ZAB were quick to accept Vinogradov into their movement and view his participation as highly beneficial to the movement. In contrast, however, specialists such as Troitsky, viewed Vinogradov as both a party infiltrator and opportunist and were highly skeptical of his purpose in participating in the movement.[20]

Throughout 1988 and 1989, the position of the oblast first secretary, Khodarev, on the GAST issue was never clear. While Vinogradov claims that Khodarev shared his views on the nuclear heating station, other evidence indicates that Khodarev hoped to be able to circumvent public opinion and bring the station on line by 1990. This was made quite clear at a session of the societal commission for control over the quality of design, construction, and operational work at Gorky AST. As reported by *Sotsialisticheskaya industriya* in April 1989, this session was supposed to be open to the public, but in fact passports were required for admission and anti-GAST activists were not permitted to attend. According to the report, only the pro-GAST viewpoint was presented at the session. The report quotes first secretary Khodarev saying that now is the time to work, not to protest, and that the petitions of the population have no relevancy in the decision on whether to allow the GAST to open. Khodarev's only concession to the population was to note that they could be assured that the station would not open until all safety concerns had been adequately addressed.

Shortly thereafter, on May 3, 1989, the Gorky oblast soviet executive committee issued another decree, number 178, simply confirming the composition of the societal commission and requesting that it continue its work up through the third quarter of the year.[21] Aside from several minimal changes in composition, this decree did not represent any radical shift in opinion among local authorities.

During the spring and summer of 1989, anti-nuclear activities were at their peak. Petition drives continued and expanded. Now, instead of a single petition table in the center of town, there were six petition stations scattered around the oblast, which collected signatures every weekend.[22] In addition, many of the factory unions voted against the completion of the Gorky AST and sent letters to authorities both in Moscow and at home.[23] Factories also threatened to strike if construction of the GAST continued, and a number of brief, two-hour strikes, were staged at various factories during the summer of 1989.[24] Both the Rossiya sovkhoz (a state-owned farm) and the all-union selection station (a seed distributor), located immediately next to the Gorky AST were quite vocal in their opposition to the station on the grounds that their products might be contaminated.

During the summer of 1989, a number of mass rallies were held throughout the oblast. During the two-month period from July to August 1989, nine rallies were held in various regions of the oblast, including Gorky, Bor, Kstovo, and Dzerzhinsky. Attendance varied, but often numbered several thousand.

Although the party attempted to present a neutral face to the anti-GAST activists during the summer of 1989, the continued opposition of much of the local party leadership to this movement was obvious. On several occasions during 1989, the umbrella group For Atomic Safety (ZAB) attempted to register as an oblast-level informal association. Their request, however, was rejected on the grounds that an anti-GAST organization already existed. In fact, the tiny group, "Women against GAST," with only five official members had petitioned for registration in 1988 and received it. Thus, the oblast executive committee argued that no other groups were needed in this issue area. ZAB's failure to receive official status during 1989 meant that the group was denied the privileges of an officially recognized association. Likhachev has noted that because of this, all funds collected by the group were simply stashed in activists' apartments, and even the existence of such funds had to be concealed from authorities.

As the anti-GAST movement grew, however, Gorky city authorities began to recognize the benefits of befriending the movement rather than antagonizing it. Thus, both the city executive committee and party organizations invited ZAB to register as a city-level organization. The group refused, however, since this would mean excluding members from other cities in the oblast.

During 1988 and early 1989, the anti-GAST activists had been pursuing a second line of demands. That is, in addition to petitioning for the cancellation of the Gorky AST, they had also petitioned for an international team of experts to visit and evaluate the station's safety. Many of the letters of protest sent by various organizations and groups to Moscow included the demand for such an international review. Obviously hoping that a positive evaluation from an international team would dispel the anti-GAST movement in Gorky, Moscow authorities acceded to this demand in March 1989. Activists were informed of this decision through letters sent out to various factories and informal organizations. Thus, for example, the director of the Borskii stekolnii zavod received a letter from the USSR Ministry of Atomic Energy and Industry dated March 14, telling him that in response to the factory's earlier demands, the Ministry had agreed to invite the International Atomic Energy Agency (IAEA) to review the Gorky station. The letter noted that a final decision would be made on the fate of the Gorky AST following the IAEA review.[25]

In early July 1989, a team of international specialists from the IAEA visited Moscow and reviewed the design of the Gorky AST. Their initial evaluation of the station was positive. Following their stay in Moscow, the team moved on to spend the entire month of August in Gorky to evaluate the station firsthand. Although representatives of the IAEA met with anti-GAST activists, the activists did not feel that their concerns were properly addressed. They were also skeptical of the impact of the IAEA review, since the IAEA could only recommend improvements, and the organization itself would bear no responsibility for the safe construction and operation of the station.

It's clear that authorities hoped that the IAEA review would calm local concerns and defuse the anti-GAST movement. While local activists were distrustful of Soviet specialists, especially the *atomshchiki* sent from Moscow, authorities hoped that an international team of specialists would carry real authority with the local population and that their positive evaluation of the station would influence popular opinion in Gorky. At a meeting of the societal commission for control over the quality of design, construction, and operational work at Gorky AST, Fedor Mitenkov, the chief designer of the station, and Mazun, the head of the safety division at GAST, proposed that the commission defer to the IAEA's judgment. The societal commission, however, was unable to reach a consensus on this point.

In November 1989, the IAEA made its findings public at an open session at the Kurchatov Institute in Moscow.[26] M. Rosen, head of the

IAEA safety subdivision, revealed that the team had found the station's design, construction, and operational plans to be quite reliable. The team noted that the lower temperatures and pressures utilized by this new design made ASTs significantly safer than normal atomic energy stations. In addition, numerous safety features which had been added to the design after the Chernobyl disaster were noted and praised. Construction quality was said to be up to international standards. In other words, the IAEA findings were unambiguously positive.

The impact of the IAEA evaluation on public opinion, however, was not what Moscow had hoped for. Many activists derided the organization as simply part of an international atomic "mafia," with vested interests in the continuation of nuclear power.[27] Others, more generously, simply noted that the international team had no understanding of construction and work conditions in the USSR and thus could not see the real dangers of the station.[28] Even the *Sovetskaya Rossiya* journalist who reported the positive findings of the IAEA commission remarked that the commission's findings were inadequate, since waste disposal, evacuation procedures, and construction flaws had not been properly dealt with by the commission. The end result was an almost-total popular rejection of the IAEA findings.

The Elections of 1990 and Resulting Competition between Levels of Authority

During the spring of 1990, parliamentary elections were held at the local, city, oblast, and republic level. As in all regions of the USSR, these competitive elections to local soviets dramatically altered the nature of public activities. Rather than focusing on public rallies and petition drives, activists in all issue areas shifted their attention first to the electoral process and then to the newly constituted soviets. Thus, the elections of 1990 provided the population with an institutional channel for funneling their demands to the government and spelled the end of the populist period of mass demonstrations and protests.

This was particularly true with regard to the nuclear power issue area. In Gorky, local candidates were only too aware of the popular appeal of the environmental and anti-GAST platform. Petition drives had already demonstrated that over 100,000 local residents opposed the station (and by some estimates, up to 200,000 people had signed petitions); thus it was obvious to all candidates that anti-GAST slogans were a powerful tool for appealing to the electorate.[29] Most people interviewed, both on the pro- and anti-nuclear side of the issue, viewed the candidates' use of the Gorky AST issue as manipulative and utilitarian, and doubted the genuine commitment of most candidates to their professed anti-GAST platform.

While environmental and nuclear power issues played a prominent role in the local elections, very few environmental activists were candidates in these elections. At the local level, Askhat Kayumov, a university student who had been highly active in the university's *druzhina* movement, was the only environmentalist to run for office. In the elections to the Russian parliament, physicist and anti-GAST activist, Boris Nemtsov was a candidate. Aside from these two exceptions, no other members of ecological groups or clubs ran in the local elections. It seems apparent from this that the shift from populist tactics to parliamentary methods was not adequately foreseen by most of the environmental and anti-GAST movement.

While most candidates may have used the Gorky AST issue in an almost entirely cynical way, the use of this issue to appeal to the electorate may very well have exacerbated fears about the station. Continued references to the possibility of "another Chernobyl almost in the heart of the city," certainly brought the issue to the attention of an even larger audience than before. Moreover, the level of discussion dropped to an even less scientific level than before, thus sensationalizing the issue and increasing public fears of nuclear power.

The results of the elections were not unexpected for such a conservative region. While a number of old party functionaries lost in the elections, the overall composition of the local soviets remained both conservative and overwhelmingly communist. Approximately 85 percent of both the Gorky city soviet and oblast soviets were made up of members of the Communist Party.[30] Of the two genuinely "green" candidates, Kayumov lost, but Nemtsov won his bid for a seat in the Russian parliament.[31]

Following the elections, the newly constituted city and oblast soviets were under considerable pressure to follow through on their electoral promises to bring about the cancellation of the Gorky AST. Thus, at the first session of the Gorky oblast soviet in early April, a rather vague and half-hearted decision was issued on the Gorky AST question. This decision, "On the construction of GAST and filling of Cheboksarsky Reservoir," was dated April 5, 1989, and stated only the soviet's support for banning further construction of GAST (as well as its recommendation that the reservoir be limited to a height of 63 meters).[32]

On April 29, the Gorky city soviet issued a somewhat more forceful decision "On halting construction of GAST."[33] In the decision, the city soviet noted that since all appeals of Gorky residents to the USSR government and the USSR Academy of Sciences to halt construction of the Gorky AST had gone unheeded, the city soviet deemed it necessary to take matters into its own hands. The document also noted that the need to halt construction on the station was based not on safety considerations but rather on psychological factors; the presence of the station in such close proximity to the city had caused great psychological stress, and the majority of residents had requested that it be canceled. Because the public did not want the station, the city soviet claimed that it would be impossible to even consider bringing the Gorky reactors on line.

The decision itself consisted of three parts: (1) that the GAST be canceled immediately; (2) that the oblast soviet take the appropriate steps to ensure that USSR authorities confirm the cancellation of GAST and halt its financing; and (3) that the Gorky city executive committee address the question of what to do with workers laid off from GAST. While the city soviet decision was certainly more emphatic and forceful than the earlier decision of the oblast soviet, it is worth noting that the city deputies recognized their own lack of authority in this issue area and appealed to the oblast soviet to take the proper steps to ensure fulfillment of the decision. In addition, the city deputies recognized the ultimate authority as lying with the USSR organs.

During 1990, the green movement underwent a fundamental transformation. Following the elections, a new environmental organization was established by Askhat Kayumov. Building largely on his connections in the university *druzhina*, Kayumov put together an organization whose primary function was to work with the city and oblast deputies in developing environmental legislation. Kayumov's organization was named Dront, and was immediately registered as an informal association. The local authorities viewed Kayumov's group as a positive step for the environmental movement, and even provided him with an office in the Komsomol building within the government complex in the Gorky Kremlin.

Ironically, it was at this point that the umbrella group, For Atomic Safety (ZAB), was finally officially registered. Because Dront was given the right to register subgroups under its organizational umbrella, ZAB was finally registered under Dront's auspices. But ZAB's day in the sun was actually over. Following the elections, green clubs whose tactics focused primarily on protests and petitions withered away. While Likhachev attempted to organize several rallies during the spring of 1990, he discovered that people were no longer interested in participating in protests and turnout was extremely low. Even the picketing of the oblast and city soviets on the days GAST decisions were expected to be issued included less than thirty people. As 1990 wore on, Likhachev and other populist activists began to recognize that the situation had shifted and they were no longer necessary. People like Kayumov who focused on parliamentary methods began to take over the environmental field.

Likhachev's final act in the anti-nuclear crusade demonstrated his recognition of the changing requirements of successful anti-nuclear activism. Almost immediately after the elections, Likhachev asked newly elected Russian deputy Yerokhin, who had been nominated by Likhachev's old factory, to set up a meeting between anti-nuclear activists and Boris Yeltsin. At that time, Yeltsin was chair of the USSR Supreme Soviet's Construction Committee, and thus played a key role in financing decisions for large construction projects. Yeltsin met with Likhachev, Gurevich, and several other ZAB members for almost an hour and was apparently sympathetic to the concerns of the group. Although he made no promises, Likhachev and others left the meeting hopeful that Yeltsin was on their side.

Meanwhile, Kayumov and his associates at Dront focused their attention on local legislation, and they were highly disappointed by the decision issued by the Gorky oblast soviet in April 1990. Thus, during the spring and summer of 1990, Dront lobbied the oblast deputies for a much stronger decision on the GAST issue.[34]

On May 7, 1990, a major step was taken toward the implementation of the oblast soviet's rather vague decree on the Gorky AST. In order to consider the options surrounding GAST, the head of the Gorky oblast executive committee, Sokolov, met with the vice chair of the USSR Council of Ministers.[35] At this high-level meeting, it was agreed that (1) the USSR Ministry of Atomic Energy and Industry would prepare a relative analysis of the technical-economic aspects of three alternative proposals regarding the fate of GAST,[36] and present their findings to the USSR Bureau of the Fuel-Power Complex, the Russian Council of Ministers, and the Gorky oblast executive committee; (2) an extradepartmental expert commission would be established to consider the impact of GAST on health and ecology and compare it to traditional power sources; (3) a second expert commission, drawing on specialists from the USSR State Committee on the Protection of Nature, the USSR State Committee for Atomic Control, the USSR State Construction Committee, the USSR Academy of Sciences, and local Gorky specialists, would be established to consider the environmental impact of GAST and to present its findings to the USSR Council of Ministers, the Russian Council of Ministers, and the Gorky oblast executive committee in 1991; and (4) the USSR Ministry of Power, the Russian Council of Ministers, Gosplan, and the Gorky oblast executive committee would jointly develop an alternative plan for supplying heat to the Gorky oblast.

On June 28, the Russian parliament took the daring step of issuing a moratorium on new nuclear construction projects in the republic.[37] This decree was viewed by most of the Gorky population as the ultimate victory in their campaign to halt the construction of the Gorky AST, and popular interest in the GAST question in Gorky dropped considerably after the moratorium was announced. In reality, however, the implications of the moratorium decree were far from clear. Although the decree was unambiguous on the question of initiating new projects, it was extremely vague on the issue of what to do with partially completed projects. Thus, while representing a symbolic victory, the moratorium did little to change the real situation in Gorky or any of the other regions where partially completed stations were being contested. Throughout the summer of 1990, neither the Russian nor the USSR government issued a decision on the fate of the Gorky AST.

During the summer, however, the societal commission for control over the quality of design, construction, and operational work at the GAST reached its final verdict on the AST. In light of the overwhelming popular opinion against the station, the commission recommended that the

Gorky AST be canceled. This recommendation, along with detailed legislative proposals prepared by the Dront staff, led the oblast soviet to reconsider the GAST question and issue a much more forceful and detailed decree in August 1990.

The August 21 oblast soviet decision (number 21), "On halting the construction of Gorky AST," was remarkable in several respects.[38] First, not only did the decree ban further construction of the AST, but it justified the oblast soviet's right to decide such questions in terms of the new USSR law on local self-rule and the new Russian law on the rights of local soviets. Such claims to the rights of oblast soviets to decide questions long considered to be the domain of the USSR authorities alone represented a radical step in the direction of local self-determination. The decree went on to demand that the USSR Council of Ministers immediately halt the financing of GAST and that the Gorky oblast executive committee develop concrete proposals for alternative usages of the GAST facilities, a plan for utilizing gas to heat the region, and a program for energy conservation.

Finally, the oblast soviet requested that their ban on further construction of the Gorky AST be confirmed by the Russian parliament and Russian Council of Ministers. This shift from USSR to Russian government organs as the ultimate authority in deciding major questions in energy planning, construction, and financing represents a dramatic shift in established procedure. Thus, for the residents of Gorky, the ultimate authority in the land had apparently shifted from the USSR to Russia.

In November 1990, the chair of the Russian Council of Ministers, Silayev, signed a decree supporting the Gorky city and oblast soviet decisions and banning all further financing of construction on the Gorky AST. At this point, the Russian and USSR authorities found themselves in a head-to-head confrontation. The USSR Council of Ministers refused to confirm the cancellation of Gorky AST and furthermore refused to halt the flow of funds to the construction project. Thus, despite the decision of the Russian government, construction at the Gorky AST continued. Interviews with the administration of Gorky AST make clear that the station director and his staff viewed themselves as employees of the USSR and thus viewed only USSR decrees as binding.

Construction of Gorky AST continued, and by early 1991, the station was basically completed. Although the USSR authorities did not dare to bring nuclear fuel to the facility, they had succeeded to a large decree in their objectives. The station was complete and ready to go into operation whenever needed. The USSR authorities were convinced that once shortages in heat began to be felt in Gorky—probably in 1992 or 1993—local residents would forget their fears of nuclear power and demand that the station be opened.

There are also numerous signs that the Gorky oblast executive committee members were similarly convinced. Although the oblast soviet demanded that the executive committee come up with an alternative plan for heating the region, there are indications that the executive committee was in no hurry to develop such a plan. In the fall of 1990, an open letter to the chairman of the oblast executive committee, Sokolov, was circulated in both the Gorky city and oblast soviets.[39] In this letter, it was claimed that the executive committee was deliberately delaying the formulation of an alternative heating plan for the region in hopes that there would be no alternative but to open the Gorky AST in 1992 or 1993. The executive committee was accused of sabotaging the oblast soviet's decision, and the letter threatened that a complaint about their inactivity might be lodged with the Russian government.

Thus, the outcome of the anti-GAST crusade was far from clear at the end of 1991. While activists had succeeded in preventing the Gorky AST from opening on schedule, their victory was far from complete. During 1991, as conditions in Russia continued to deteriorate, interest in environmental issues and the nuclear power question eroded rapidly. Not only did activism in these movements fall to insignificant levels but local legislators who had vehemently opposed the Gorky AST in the spring of 1990 began to waver on the issue. Without the pressure of overwhelming public opinion, the incentives to eventually reverse the Gorky AST decision and to reap the huge savings in supplying heat to the region with nuclear power appeared high.

Interestingly enough, however, with the dissolution of the USSR in 1991 and the achievement of genuine Russian independence, the question of the fate of the Gorky AST was finally resolved. In keeping with Russia's moratorium on nuclear expansion as well as the introduction of more rigorous restrictions on the siting of nuclear facilities, plans to open the Gorky AST were indefinitely tabled. Because of the station's experimental design and its unusually close proximity to a highly populated area,

Russia's Atomic Control Commission (Gosatomnadzor) recommended against opening the facility.[40]

Russia: *The Missing Nationalist Link*

One of the most curious features of the anti-nuclear movements of Russia is their lack of identification with the Russian nationalist cause. If we were to construct a spectrum running from one extreme of complete overlap between nationalist and anti-nuclear aspirations to the other extreme of total separation of the two platforms, we would find Lithuania and Armenia very close to the former extreme, Russia at the other end of the spectrum, and Ukraine somewhere in the middle.

In Russia, anti-nuclear activists were fighting not only against a particular nuclear power station but also for greater local decision-making rights. As in the other republics, the anti-AES movements were not purely environmental but incorporated an important political component. Thus, Russian activists consistently voiced their mistrust of Moscow decision makers and their hostility toward Moscow's interference in their daily lives. In contrast to the other republics, however, activists were unable to translate this anti-Moscow feeling into a pro-Russia movement.

I would suggest that the reason for this distinctive feature of the Russian anti-nuclear movements lies with the unique position of Russia within the USSR. While Lithuania could decry Moscow's intervention in their republic affairs and deride the USSR as simply an occupying force, the Russians found it extremely difficult to paint their situation since 1917 as one of foreign occupation. After all, it had been the "Russian" revolution which brought the Soviet government to power and created the institutions of state that were now becoming ever less acceptable to Russia's citizens. Whereas other republics had a clear enemy to mobilize against—the USSR—Russia did not. The Russian and Soviet national identities were so intertwined that it was extremely difficult to mobilize Russian society against a common enemy.

While other republics were able to unify their people against a clear enemy, Russian nationalists had to find other grounds for mutual identification. Thus, the Russian nationalists of the early perestroika period tended to search back to their old ethnic Russian roots to find a national identity around which to unify. This meant that Russian nationalism initially tended to be quite chauvinistic, based on a sense of ethnic identity and common language, religion, and cultural experiences. The Pamyat movement is the best example of such a chauvinistic and ethnically oriented Russian nationalist movement.

Interestingly enough, however, an exclusive ethnic identity proved to have little appeal to the population of Russia in the 1980s. As the most multi-ethnic of all the republics, an exclusive definition of the nation based on ethnic heritage not only excluded huge sectors of the population but seemed to many citizens to be totally inappropriate to the modern era. The Russian ethnic identity that Pamyat was attempting to resuscitate seemed more appropriate to nineteenth-century tsarist Russia than to the modern, multi-ethnic republic of the late-twentieth century. Support for Pamyat and other ethnically based Russian nationalist groups was thus extremely weak (despite the exaggerated fears of the Western press in the late 1980s).

Thus, the anti-nuclear movements that emerged in 1987–89 in Russia were political to the extent that they were a cover for demands for more local decision-making rights, but they were neither a surrogate for Russian nationalism nor even remotely intertwined with the ethnically based nationalist groups.

In 1990, however, a slight nationalist element crept into the anti-nuclear movements with the ascendancy of Boris Yeltsin. While prior to this time, no attempts had ever been made to distinguish Russia from the USSR, Yeltsin startled the whole world by suddenly demanding Russian sovereignty from the Soviet Union. In an extremely bold and creative move, Yeltsin announced that all residents of the Russian territory were Russian citizens (rossiyanin), and they deserved sovereignty and independence as much as any of the other republics of the USSR. This redefinition of what it meant to be a member of the Russian nation radically reshaped Russian nationalism and suddenly allowed the people of Russia to unify against a common enemy, much as had occurred in the other republics.

Thus, very late in the game, just as the anti-nuclear movements were fading away, a weak nationalist linkage could be detected. In 1990, following Yeltsin's redefinition of Russia's relationship to the USSR, anti-nuclear activists reoriented their tactics and suddenly began to appeal to the Russian government to confirm their anti-nuclear decisions. Thus, anti-nuclear activists looked to the civically defined Russian nation-state as the ultimate authority in the land. While the anti-nuclear movements were neither a surrogate for Russian nationalism nor intertwined with nationalist movements, in the end they supported Yeltsin's definition of the Russian nation and his demands for the creation of a Russian nation-state.

CHAPTER 6

The National Enclaves: Tatarstan and Crimea

While the boundaries of the fifteen republics of the USSR provided the focus for many of the most prominent independence movements of the perestroika period, the drive for national autonomy was not limited to the titular national groups of the fifteen republics. The Soviet Union was composed of over 100 different national groups, many with substantial populations and well-defined territorial boundaries. In fact, thirty nationalities had populations between 100,000 and 1,000,000, while twenty-two more possessed populations exceeding 1,000,000. Clearly, the urge for national self-determination was unlikely to be fulfilled merely by the dissociation of the USSR into its fifteen constituent republics.

As was the case with the titular national groups of the republics, the drive for national autonomy by minority nationalities was fueled by the long-standing administrative boundaries of the former USSR. Each republic was divided into a number of territorial subunits, including autonomous soviet socialist republics, *krais,* and oblasts. In many cases, these administrative boundaries again overlapped with those of the dominant nationality in the region. Thus, those minority nationalities within each republic with a feasible claim to a subunit of the republic often rejected the independence goals of the republic's titular nationality and instead pursued their own objectives of national autonomy. This tendency, which was first observed prior to the dissolution of the USSR, has continued to plague many of the newly independent states of the former Soviet Union. In this chapter, we will consider the linkage between national sovereignty and anti-nuclear movements in two particularly important national enclaves: Tatarstan (Russia) and Crimea (Ukraine).

In Russia, the republic's multinational composition has provided ample opportunity for enclaves of minority nationalities to pursue their own political objectives. Within the Russian Federation, twenty-one autonomous soviet socialist republics existed whose boundaries had earlier been established according to ethnic principles. To many, these ethnically delineated republics continue to represent a powder keg for Russia, possibly setting the stage for the Russian Federation to replicate the path of the USSR and to fragment along ethnic lines. While most of these republics have yet to openly challenge Russia's authority, the republics of Chechnya and Tatarstan have proven particularly troublesome for Moscow. In late 1991, Chechen nationalists unilaterally dissolved the Chechnya-Ingushetia Autonomous Soviet Socialist Republic (ASSR) and proclaimed the birth of an independent Republic of Chechnya. Similarly, in March 1992 the population of Tatarstan passed a referendum declaring their republic a sovereign state whose relations with Russia should be regulated by treaties between equal partners. While the Chechens ultimately failed to resolve their differences with the Russian Federation and have suffered disastrous consequences, the Tatars have made significant progress in normalizing relations with Moscow and have recently concluded a bilateral treaty regulating key transactions between Russia and Tatarstan. The treaty, however, stops short of recognizing Tatar sovereignty or independence and it is unclear whether this measure will ultimately resolve the standoff between Moscow and Tatarstan's capital city, Kazan.[1]

Because Tatarstan was targeted for a new nuclear power station in the 1980s, anti-nuclear mobilization once again provides a lens through which to view the emergence of nationalism in the region. As elsewhere, the nuclear power issue was one of the first to mobilize the masses in Tatarstan, and its linkage to nationalist mobilization in the republic provides insight into the strength and nature of the Tatar national identity. How potent was Tatar nationalism after four centuries of domination by Russia? Was the anti-nuclear movement simply a surrogate for Tatar nationalist objectives? Or did centuries of Russification blur the boundaries between Tatar and Russian identities and lead to the confusion in defining national boundaries observed in Ukraine?

The ethnic situation in Ukraine is far simpler than the situation in Russia. While pockets of non-Ukrainian ethnic minorities do exist, most minority groups are small, politically weak, and lack clearly defined preexisting territorial boundaries to assist in the mobilization process. The situation of the largest ethnic minority in Ukraine—the Russians—however, may have the potential for intense interethnic conflict within the newly independent country of Ukraine. Because Russians, which make up almost one quarter of Ukraine's population, are dispersed across most of the country, at first glance, opportunities for Russian separatism in most of Ukraine appear slim. The situation in the Crimean region of Ukraine, however, does provide cause for concern. Because Crimea was only recently (1954) transferred from Russian to Ukrainian jurisdiction, has an overwhelming Russian majority, and has been closely associated with Russia (not Ukraine) for several centuries, the potential for the emergence of a Russian national independence movement in Crimea during the perestroika period and after has been substantial.

As in Tatarstan, anti-nuclear protest represented the first instance of mass mobilization in Crimea. By the late 1980s, the Crimean nuclear power station was nearing completion and preparing to go into operation; it presented a clear target for budding activists on the Crimean peninsula. Whether the mass movement that emerged was linked to nationalist goals in Crimea, however, was not immediately clear. Was the anti-nuclear movement simply a surrogate for Russian nationalism? Or was it perhaps a shield for the growth of a Crimean Tatar national movement—a smaller minority group also present on the Crimean peninsula? Or, like its counterparts in the Russian regions of the Russian Federation, was the anti-nuclear movement in Crimea largely dissociated from nationalist mobilization?

The linkages between anti-nuclear activism and nationalist mobilization which are revealed in the enclaves of Tatarstan and Crimea, display subtle differences. While the anti-nuclear movement did provide the context through which long-dormant Tatar national identities could be revived and reborn in Tatarstan, the movement played a far less important role in nationalist mobilization in Crimea. As with the cases in preceding chapters, this variation in the linkage between anti-nuclear mobilization and nationalism highlights key differences in the strength and identities of the national communities present in these regions.

Tatarstan: *The Resurgence of the Dormant Tatar Nation*

The territory formerly known as the Tatar ASSR is located in central Russia, with its capital city of Kazan well situated along the Volga River. The region is home to the Volga Tatars, a Turkic, Muslim group that traces its heritage in the region back for centuries. While the origin of the Volga Tatars is still debated—some historians trace their introduction to the region back to the Golden Horde which swept through central Russia in the thirteenth century while others argue that their roots lie with the Volga Bulgars who first settled in the region in the eighth century—no one disputes the Volga Tatars' strong historical claim to the land along the Volga.[2] Since the conquest of the region by Ivan the Terrible in 1552, however, the Tatars have been under continuous Russian domination and, as a result, have repeatedly attempted to break free of Moscow's tutelage and to reassert their independence.

By 1917, a strong Tatar nationalist movement determined to throw off the yoke of Russian imperialism had emerged. In the civil war that followed the Bolshevik seizure of power in October 1917, however, the Tatars found it necessary to choose sides and were enticed over to the Bolsheviks by grandiose promises of greater autonomy for the Muslim peoples of the new union.[3] As the civil war began to wind down, two autonomous soviet socialist republics were created in the region: Bashkir ASSR (1919) and Tatar ASSR (1920). The boundaries of the two autonomous republics were, however, drawn without regard for the ethnic heritage of the region: more than three quarters of the Tatar population was left outside the borders of the Tatar ASSR, while Tatars in fact made up the largest ethnic group in the new autonomous republic of Bashkir. The poor overlap between the boundaries of the Tatar ASSR and the population distribution of Tatars in Russia has long complicated the Tatars' struggle to establish a Tatar nation-state and has continued to be an important factor in Tatar-Russian relations to this day.

Throughout the seventy-year history of the USSR, the Tatars repeatedly demanded that their position in the union be upgraded to that of a union republic. With the founding of the USSR in 1922, the Tatars were granted only the status of an autonomous republic within the Russian Republic—a component of Russia rather than an equal. Their request for union status, first in 1922 and later with the adoption of the constitution of 1936, was consistently denied. Periodically resurfacing in the 1960s and 1970s, the issue of the Tatar ASSR's status as a subunit of Russia has been a continuous source of irritation for the Tatars. As Tatars make up the second-largest ethnic group in Russia (behind the Russians themselves) and the fifth-largest in the entire USSR (after the Russians, Ukrainians, Uzbeks, and Belarussians), their claim to greater rights to union republic status than the Kirgiz, Balts, etc., was not without foundation.

By 1989, the ethnic composition of the Tatar ASSR had changed dramatically since the days before the Russian conquest of 1552. Because the Tatar territory represented such a key strategic region—opening a gateway to Siberia, providing access to the Caspian Sea, and controlling trade routes along the Volga—its capital city of Kazan quickly grew into a major Russian trading center. Russians flooded into the region, thus beginning the shift in ethnic composition that was to continue well into the twentieth century. During the Soviet period, Tatar ASSR became a major industrial center for the USSR, with a high concentration of heavy-industry, military-production, and oil-extraction facilities. Once again this implied an immense inflow of Russians to tend to the growing industrial sector. The result of this continuous Russian migration was the dilution of the Tatar population until, by 1989, Tatars represented only 48.5 percent of the population of Tatar ASSR (43.3 percent of the republic was ethnically Russian).[4]

Over the past several centuries, the Volga Tatars have struggled to resist the almost overwhelming pressures toward Russification. Despite harsh persecution of their Muslim traditions and faith, particularly during the Stalin period, the Tatars have managed to maintain a strong religious orientation. Their attempts to preserve Tatar language and culture against the onslaught of Russification, however, have been less successful. Their indigenous language and culture was even less protected than the languages and cultures of the fifteen union republics. National elites were granted fewer opportunities for upward mobility, and by the late 1980s, Tatar culture was viewed by many as in danger of extinction. During the 1960s, Tatar language schools had been completely eliminated, and the older generation of Tatars feared that the Tatar language would die with them. Due to the heavy influx of Russians to the cities of the Tatar ASSR, urban Tatars were especially vulnerable to Russification. Intermarriage was common between Russians and Tatars, and by the late 1980s, the Tatars' sense of a distinctive national identity was clearly threatened.

The ethnic composition of the Tatar ASSR, split almost evenly between Tatars and Russians, along with the long history of Russification in the region, made the potential for nationalist mobilization in the region uncertain. Would the introduction of perestroika mean the growth of a unified Russian-Tatar front, or would these new freedoms of expression and action bring the rebirth of a long-dormant sense of Tatar national identity? In looking at the Tatar case, it is immediately obvious that the population of Tatar ASSR lacked the unified sense of national identity and history that characterized the Lithuanians in the late 1980s. If mobilization was to occur along ethnic lines, it would require the resuscitation of an endangered sense of nation and a slow, arduous process of redefining national identity. Unlike the Ukrainian case, however, the languages, cultures, and religious traditions of the two dominant ethnic groups were quite distinctive, and people were unlikely to suffer the confusion of national identity that occurred between Russians and Ukrainians in Ukraine. Thus, the potential for a growing nationalist schism in the population of Tatar ASSR was clearly present, and it only remained to be seen whether nationalist political entrepreneurs would be able to play upon ethnic differences and successfully trigger mobilization along ethnic lines.

Social Mobilization in the Tatar ASSR

As elsewhere in the Soviet Union, social activism in Tatar ASSR began with the relatively "safe" issues of ecology, culture, and history. In 1987 and 1988, small discussion clubs began to coalesce, and the first tentative steps were made toward mass mobilization. Early mass actions were linked almost exclusively to environmental questions. Heavy industrialization of the republic had created a myriad of environmental problems which had long gone unchecked. Extensive contamination of the air, water, and soil of the Tatar ASSR made the republic ripe for the emergence of a strong environmental movement.

During 1987 and 1988, government plans to construct buildings or production facilities in wooded or recreation areas formed the focal point for Kazan's early environmental activists. Students and faculty at the republic's prestigious Kazan State University provided the initial leadership in mobilizing the population to oppose unwanted development projects. Mobilization began slowly, with the first issue tightly linked to the interests of the Kazan State University community. In 1987, it was learned that the government planned to raze a popular wooded area contiguous with the university in order to make way for a construction project. Students and faculty associated with the city's most active *druzhina* (an official student environmental club), in the university's biology department, strongly opposed the project and took the bold step of contesting the government's decision. Up until this time, the *druzhina* had engaged in the normal, accepted campaigns of all official *druzhina* of the USSR: primarily, the crusades against poaching and illegal Christmas tree acquisitions. Now, the *druzhina* decided to take on the authorities.

Interest in the issue grew, and soon much of the university community was involved along with residents of the region slated for development. Kazan's first popular protest, a picketing action held near the university, was unexpectedly successful. Attendance was substantial and the city's most progressive newspaper *Vechernaya Kazan* provided extensive, positive coverage of the event. In this first demonstration of the power of the people, the government was quick to back down and agree to relocate their planned construction project.

Building on this early success, the next campaign undertaken by Kazan's budding environmental activists was more ambitious. Moscow planners had targeted Kazan for the construction of a new biochemical factory which was to produce the controversial livestock supplement BVK. During 1987 and 1988, BVK factories across the USSR became central targets for environmental protest due to the dangerous air pollution associated with the production of BVKs as well as the fact that BVKs were banned in much of the developed world. Plans to build a BVK factory in Peschaniye Kovaly, a recreational area on the outskirts of Kazan, angered many sectors of Kazan's population and brought about environmental mobilization on a much broader scale than the earlier crusade.

During the spring of 1987, the planned BVK factory became a hot topic of discussion among Kazan's population. *Vechernaya Kazan* provided a forum for the debate which occurred frequently in the pages of this newspaper during January through June 1987.[5] Once again, the faculty of Kazan State University took a leading role in the campaign, frequently publishing attacks on the factory plans and holding open forums to acquaint the population with the perceived dangers of such a facility.[6] Petition drives were undertaken and the newspaper *Vechernaya Kazan* reported having received over 50,000 letters opposing the factory by mid-1987.[7]

With the campaign against the BVK factory, participation in the environmental movement began to broaden. Concerned citizens outside the university community began to take active roles in organizing mass actions, and a broader environmental network began to coalesce. The first independent environmental club in the city, Ekologicheskii klub, emerged at this time. Without a tight organizational structure, regular meetings, facilities, and other attributes of established environmental associations, this club still represented an important step in the future organization of environmental activism in Kazan. Other small, local groups also emerged at this time to assist in the crusade against the factory. The fledgling movement held several mass protests in Kazan and is attributed with bringing about the first genuine mass mobilization in the city. Participation in protest rallies often numbered in the thousands—a milestone for Kazan.

In response to this unexpected outburst of popular opposition, local authorities were quick to join the protestors and demand the cancellation of the project.[8] The Tatar Council of Ministers officially requested that USSR authorities reconsider the siting of the factory, and in July 1988, USSR Gosplan acceded to local demands—once again reinforcing the peoples' sense of efficacy in their actions.[9]

While other environmental actions against planned industrial facilities followed in 1988 and later,[10] none were as ambitious as the battle against the Tatar Atomic Energy Station (TAES). While most of these early environmental campaigns targeted facilities yet to be constructed (thus lowering the cost of cancellation), the anti-TAES campaign focused on an immense construction project into which millions of rubles had already been sunk. With construction having begun in 1983,[11] the station was well on the way to being operational by the time opposition to the AES was first voiced by Kazan activists in 1988. Over half a billion rubles had already been spent, and the first reactor was scheduled to come on line in 1990, just two short years away.[12] While the station was located over 100 kilometers from Kazan in the town of Kamskaya Polyana, environmental activists in the capital city were the first to raise the cry against the republic's first and only nuclear power station.

Anti-nuclear activists in Kazan can be separated into several distinct strands: specialists, students, and the broader populace. Among scientific specialists, the issue of the Tatar AES became a matter of concern in early 1988. Once again, academics at Kazan State University provided the most active core of specialist opposition, and the city's progressive newspaper, *Vechernaya Kazan,* provided the forum for discussion. Focusing mainly on technical deficiencies, such as the siting of the station in a seismic zone and inadequate water sources for cooling, numerous well-respected specialists publicized their concerns through both the press and open discussion sessions. While these specialists never organized their opposition into a formal club or association nor joined the mass movement which was emerging at this time, they nonetheless played a critical role in the anti-nuclear campaign. Scientists of significant stature, including Yurii Kotov, Chairman of the Ecology Department and Dean at Kazan State University; Boris Burov, Chairman of the Geology Department; and A. Konovalev, a prominent biologist at the university, provided the emerging movement with the technical arguments needed to wage a successful campaign and thus indirectly lent the emerging movement scientific legitimacy and credibility.[13]

The Kazan State University student *druzhina,* which had begun to reject its role as a docile and obedient official club with its participation in the environmental campaigns of 1987–88, was also quick to join in the anti-TAES crusade. With the stakes being so high, the anti-TAES campaign was viewed as highly political and controversial from the start, and the participation of the *druzhina* in such a confrontational issue was viewed by many as a radical departure from the accepted mission of these officially sponsored organizations. While the *druzhina* at Kazan State University and the other many *druzhinas* at institutes around Kazan had long been considered in decline, suffering from overwhelming student apathy and indifference, the campaign against the Tatar AES injected these student organizations with a new dynamism and sense of mission. Many student participants, however, hasten to add that although the anti-TAES battle reinvigorated a dying movement, students never became the leading or dominant force in the anti-nuclear power movement in Kazan.[14]

Interestingly enough, the dominant force in the mass mobilization of the anti-TAES movement was neither scientists nor students, but rather one extremely committed individual named Albert Garapov. Garapov was an engineer with no professional linkage to the nuclear power question who almost single-handedly organized the anti-nuclear activities which gave this movement its mass character. Throughout the anti-nuclear struggle, which lasted from 1988 through 1990, Garapov could always be found at the center of all mass activities. Garapov, however, worked poorly in groups, and thus his leadership of the movement was only marginally connected with any organized associations. While he often mentioned his leadership of the Anti-Nuclear Society, this supposed association had no regular meetings, meeting space, or organizational structure. When an informal environmental newsletter began to be published in the fall of 1989, the name of the editor (and probable writer) was never included, but the phone number of the editorial office was (not surprisingly) Garapov's home telephone number![15] Garapov's energy and dedication to the tasks of organizing the movement were prodigious, with his constant publication of anti-TAES articles in the press, publication of his own newsletter, circulation of petitions, and organization of mass actions. It is probably most accurate to characterize this dominant force in the anti-TAES movement as a small and constantly changing group of friends and acquaintances which revolved around a single committed individual, Albert Garapov.[16]

The actual crusade against the Tatar AES began in earnest in 1988. During the spring of 1988, articles by specialists and by a very active journalist at *Vechernaya Kazan,* Gennadi Naumov, began to introduce the public to the potential dangers of the almost completed nuclear power station to the north of Kazan. During the summer of 1988, the first mass protest rally against the station was held in Kazan. Outside the capital city, however, society remained largely quiet, with little sign of either ecological or political awakening.

It was not until 1989 that the movement began to spread across the Tatar ASSR. The elections for the USSR Congress of People's Deputies provided an opportunity for the anti-nuclear issue to receive widespread attention, and as in other regions of the USSR, opposition to the local nuclear power station was adopted as a key component in almost all electoral platforms. Demonstrating the widespread popularity of this issue, leading opponents of the Tatar AES proved highly successful in the elections. Prominent anti-nuclear scientist, A. Konovalev, and the dean and chairman of the ecology department of Kazan State University, Yurii Kotov, were both elected to the Congress of People's Deputies. Another long-time activist often associated with Garapov's activities, A. Gavrilov, also won a seat in the Congress. Unlike many others across the USSR who were elected on anti-nuclear platforms, all three continued to promote their anti-nuclear views from their new positions in Moscow.[17]

Soon thereafter, mass anti-nuclear activities began to be observed and promoted outside the city of Kazan. In April 1989, a small anti-TAES rally was reported in Nizhnykamsk, an industrial center approximately fifty kilometers from the nuclear station.[18] Two weeks later, a march was conducted from Kazan to the station.[19] Activists met at Kazan State University and were shuttled by bus across the countryside toward the AES site in Kamskaya Polyana. In each town, the activists disembarked and attempted to hold local rallies to publicize their concerns about the growing Tatar station. While certainly generating a great deal of publicity for the anti-TAES campaign, it is clear that the Kazan activists were not entirely successful in mobilizing the rural countryside. In many towns, local authorities opposed the rallies and went as far as preventing the activists from disembarking within town limits. Furthermore, local turnout at the rallies was often smaller than expected by the activists.[20] Nonetheless, the action was the first of its kind to be held in the USSR and was clearly important in the early activation of the Tatar countryside.

During the remainder of 1989, the Tatar AES continued to be a hot issue in the Tatar ASSR. The conclusions of a scientific commission established by the USSR Academy of Sciences confirming the central government's claims to the safety of the Tatar AES only seemed to fuel anti-nuclear and growing anti-Moscow sentiment in the autonomous republic.[21] During the summer of 1989, attempts were made to resolve the differences between Kazan's scientific community and the Moscow specialists, but with little success. Reports of these scientific roundtables indicate that the discussions were emotionally charged and largely unproductive.[22]

By late summer, however, a dramatic change in the attitude of the Tatar authorities began to become evident. Whereas anti-nuclear activists complained of government obstruction and official opposition to their activities during 1988 and early 1989, by mid-1989, authorities began to take a new view of the issue. Recognizing the popularity of the anti-nuclear platform[23] (as well as the need to woo the electorate for the impending 1990 local elections), Tatar republic authorities dropped their opposition to the movement and began to take steps to actively demonstrate their support for the anti-TAES demands. After attending a roundtable of Moscow and Tatar specialists during the summer of 1989, Yurii Voronin, Chairman of the Tatar ASSR Gosplan and Deputy Chairman of the Tatar ASSR Council of Ministers, announced that the Tatar government's position was in concordance with the scientists'; since the Tatar scientific community recommended halting construction on the new nuclear station, the Tatar government would henceforth take steps to encourage Moscow decision makers to halt the project.[24]

During the fall of 1989, anti-nuclear activists continued to pressure Moscow to halt construction of the station. Activists staged a second march to the Tatar AES which set out from Kazan on September 30, 1989.[25] In a controversial step, march organizers included a threat of a preliminary regional strike in the march resolutions. Interestingly enough, the Kazan specialist community viewed the strike threat as overly antagonistic and counterproductive, and both Konovalev and Burov joined republic authorities in a televised appeal to cancel the strike. In hopes of resolving the growing conflict between Moscow and the Tatar republic, a committee of specialists, including Konovalev and Burov, were immediately dispatched to Moscow to lobby the USSR Council of Ministers to accede to the demands of the Tatar authorities.[26]

Moscow authorities, however, showed little willingness to accede to local demands to cancel the almost completed station. In October, Deputy Chairman of the USSR Council of Ministers Lev Ryabev publicly reaffirmed the government's commitment to open the new AES.[27] The appeals of the Tatar specialist committee which visited Moscow in October also fell on deaf ears; in a telegram to the republic's green activists, Deputy Minister of Atomic Energy A. L. Lapshin stated that additional research had found no serious geological problems associated with the Tatar AES and reaffirmed that the station would open as planned.[28]

The struggle between the two levels of authority—Moscow and Kazan—continued unabated through 1989 and early 1990. On November 4, 1989, the Tatar Supreme Soviet issued a decree calling on the USSR Council of Ministers and USSR Supreme Soviet to cancel the Tatar AES on the grounds of poor technological planning. The decree further called on the Tatar Council of Ministers to begin to lay the groundwork for converting the station to a conventional power facility or other industrial function.[29] In voting for the cancellation of the station, Supreme Soviet deputies were no doubt considering their electoral prospects in the upcoming March elections. By this time, opposition to the Tatar AES was overwhelming, with petitions demonstrating widespread support for cancellation and surveys indicating that as much as 90 percent of the population of Tatar ASSR favored halting construction of the AES.[30]

As expected, opposition to the Tatar AES was a prominent component of almost all successful electoral platforms in March 1990. The new Supreme Soviet thus came to office with a clear mandate to block further work on the Tatar AES. Thus, during the first session of the new parliament, deputies voted unanimously to halt construction and funding of the station.[31] The cancellation of funding for the station, however, was largely symbolic, since construction was ultimately financed by USSR organs, and construction on the station continued unabated during the spring and summer of 1990. In frustration, Tatarstan's greens held yet another march from Kazan, this time crossing over into Bashkir ASSR to protest the Bashkir AES as well.[32]

Interestingly enough, it was not Tatarstan's demand for cancellation of the station that ultimately forced the USSR ministries to accede to popular demands. The resolution of the issue came only with the growing power struggle between Moscow's two levels of authority: USSR and Russia. As Boris Yeltsin began to assert Russia's authority against the USSR's, cancellation of unwanted nuclear facilities became a symbol of Russian as well as Tatar sovereignty. Thus, following Russia's declaration of sovereignty, the Russian Supreme Soviet moved quickly to pass a moratorium on the construction of new nuclear facilities in the Russian Federation.

Since the Tatar AES was more than half completed, the moratorium left its fate somewhat ambiguous. During the fall of 1990, however, USSR authorities agreed to give broad interpretation to the Russian moratorium and halt construction on most nuclear facilities in the Russian Federation, whether in the early or late stages of construction. In an October meeting attended by Lev Ryabev, deputy chairman of the USSR Council of Ministers; Yurii K. Semenov, USSR minister of power and electrification; Reshetnikov, USSR deputy minister of atomic energy and industry; and Yurii

Voronin, deputy chairman of the Tatar Council of Ministers, USSR authorities finally agreed to cancel the Tatar AES and to convert the facility to a conventional heating station.

With the Russian moratorium and its confirmation by USSR authorities, the anti-nuclear power movement in the Tatar ASSR quickly withered away. In fact, environmental activism of all kinds virtually disappeared from the scene in late 1990. The few remaining environmental activists who could be found in 1991, all reported dismay and disillusionment with the rapid unraveling of the environmental movement in Tatarstan. As the chairman of the Ecology Committee of the Tatar Supreme Soviet, Aleksei Kolesnik, noted, people were beginning to expect the new official environmental organs to take care of such issues and attention was quickly turning to the now overriding concern for Tatar sovereignty and independence.[33]

Nationalism and the Anti-nuclear Crusade

As might be expected, the environmental movement did not emerge in isolation in the Tatar ASSR. As elsewhere, mobilization on environmental, cultural, and other seemingly apolitical issues, led quickly to political mobilization. By 1988, a myriad of informal groups were beginning to proliferate across the republic, particularly in the capital city of Kazan.[34] In June 1988, the budding political activists of Kazan came together to form the Tatar People's Front, an umbrella group dedicated to furthering the reform goals of perestroika.[35] The initiative group was composed of approximately fifty activists and represented the coalescence of virtually all active forces in Kazan. As expected, leading environmental activists such as Albert Garapov, played a key role in the creation of the Tatar People's Front, and the campaign against the Tatar AES was one of the first activities of the newly formed movement.

Initially, the Tatar People's Front (TPF) was not affiliated with any particular ethnic group. It was an umbrella that brought together both Russians and Tatars demanding political and economic reform. In addition to supporting perestroika, however, Tatarstan's new political activists also had their own agenda—promoting the sovereignty of the Tatar ASSR and upgrading its status to that of a union republic, on par with Russia,

Ukraine, and other union republics. Thus the new movement initially blended popular demands to cancel the Tatar AES with new demands for the greater self-determination of the autonomous republic.

During 1988 and 1989, the Tatar People's Front was actively involved in organizing anti-nuclear activities in the republic. Upon its formation, a special committee was established to direct anti-nuclear activities. In reports of the marches and demonstrations that were held in late 1988 and 1989, the Tatar People's Front is consistently mentioned as a key action organizer.[36] In addition, the newsletter of the TPF, *Atmoda,* frequently included discussion of the Tatar AES issue and plans for upcoming anti-TAES activities.[37]

While the Tatar People's Front was created to unify all active reform forces in the Tatar ASSR, the union between Russians and Tatars proved short-lived. During 1988, steps were taken toward the creation of a new political organization, the Tatar Public Center (TOTS).[38] With the creation of TOTS, Tatar participation within the Tatar People's Front began to decline as Tatars shifted their membership from the broader TPF umbrella to a more ethnically based organization. While the TPF was dedicated to further the goals of political and economic reform and work to enhance the status of Tatarstan within the USSR, the TOTS focused more attention on the revival of Tatar language and culture and the eventual achievement of a sovereign Tatar nation.

Like the Tatar People's Front, TOTS also moved quickly to establish a committee on the Tatar AES question. Thus, during 1988, Kazan's most active opponent of nuclear power, Albert Garapov, found his organizational affiliation constantly changing. While he initially claimed affiliation with the somewhat nebulous Ekologicheskii klub, in June of 1988 Garapov participated in the founding of the Tatar People's Front and became a leading player in the TPF anti-nuclear committee. With the formation of the Tatar Public Center, however, Garapov once again shifted his organizational ties and joined the TOTS committee on the TAES. Finally, however, both the TOTS leadership and Garapov himself acknowledged that Garapov's single-minded dedication to halting the construction of the Tatar AES and his inability to work with others in planning anti-nuclear activities made him a poor committee member. By the end of 1988, Garapov had quit the TOTS committee on the TAES and established his own organization, the Anti-Nuclear Society. With no formal organizational structure, regular meetings, or established membership, the Anti-Nuclear Society was unabashedly a front for Garapov's personal anti-nuclear activities.

In breaking away from the TOTS committee on the TAES, however, Garapov did not relinquish his secondary goal of rebuilding the Tatar nation. As a vehement Tatar nationalist, Garapov consistently viewed anti-nuclear activities in the ASSR as a symbol of the struggle for Tatar sovereignty and independence. His discussions on the evils of nuclear power frequently included warnings of the threat to the Tatar nation. Rebuilding a sense of Tatar identity and protecting the Tatar people from the dangerous nuclear policies of Moscow went hand-in-hand for Albert Garapov.

It is clear that the anti-nuclear movement in Tatarstan played an important role in the political mobilization of society. As the first focus for mass mobilization in the Tatar ASSR, the issue represented a key with which to unlock the pent-up political aspirations of the population. Marches through the countryside in 1988 and 1989 were an effective tool for activating the passive rural population of the republic. Initially, however, the movement was not focused on ethnic mobilization. Russians and Tatars alike were encouraged to fight against Moscow's arrogant treatment of the Tatar ASSR and to support the growing drive for increased sovereignty and status for the region. Initially, all ethnic groups were called upon to join the struggle to provide the Tatar ASSR with its rightful status as a full union republic.

It was not until 1989 that the anti-nuclear movement underwent a subtle transformation; the split between ethnic Russians and Tatars was beginning to emerge. As the year progressed, the long-suppressed distinctions between the two ethnic groups began to become more apparent. Dedicated to reversing the Russification of the Tatar population, the Tatar Public Center grew quickly in popularity and membership. While the Tatar People's Front seemed to be struggling to identify its mission and constituency during this period, TOTS was growing into a vibrant organization with a deep sense of its popular charge.

By 1990, the potential for a dangerous ethnic cleavage in Tatarstan's population began to become apparent. As political mobilization began to split along ethnic lines, with Tatars favoring the Tatar Public Center and Russians drawn to the Tatar People's Front, many feared that the autonomous republic was on the road toward violent ethnic conflict. During this time, radical Tatar nationalist organizations began to emerge, calling for the creation of an ethnic Tatar nation-state. Groups such as Ittifak were determined to reverse the centuries of Russification inflicted on their people and to create an exclusive nation-state open only to ethnic Tatars.

This radicalization of the Tatar national movement, however, did not represent a dominant trend. In fact, both of the leading political organizations—the Tatar Public Center and the Tatar People's Front—strongly opposed ethnic exclusivity and supported a civic definition of the emerging nation. Both groups went to great lengths to ensure that their organizational programs permitted membership for all ethnic groups and explicitly favored equal treatment for Tatars and Russians alike. The Tatar Public Center's civic orientation was particularly important in avoiding the erosion of ethnic relations in Tatarstan. Because Tatars had a stronger history of resistance to Moscow's domination and were able to appeal to the population on the basis of both ethnic identity and instrumentality, the TOTs quickly grew into the most popular and influential political organization in the autonomous republic, rapidly dwarfing the Tatar People's Front. Thus, it was particularly important that the Tatar Public Center adopt an inclusive approach toward the ethnic Russian population of the autonomous republic.

The dominance of a civic over an ethnic definition of the Tatar nation may be explained by several factors. First, as noted in earlier chapters, both the history and the character of national identity played key roles in shaping the Tatar national movement. Due to Tatarstan's centuries of affiliation with Russia, the Tatars' sense of a distinctive national identity had been significantly eroded. Russification of language and culture as well as high levels of intermarriage between the two ethnic groups tended to ameliorate the potential for violent confrontation between Tatars and Russians. The Tatar population lacked a strong sense of their distinctive national identity and thus were slow to mobilize on the basis of ethnic exclusiveness.

In addition to the obstacles to rapid Tatar nationalist mobilization, a shift toward an inclusive definition of the nation was also favored by the attitudes of Tatarstan's Russian population on the questions of sovereignty and independence. Due to Tatarstan's well-known success as an industrial center of the Soviet Union, calls for greater republic autonomy were strongly supported by the Russian as well as the Tatar population. In fact, both groups were united in their perception that both Russia and the USSR were draining Tatarstan of its wealth and resources and that sovereignty or independence would be economically advantageous to the republic. Thus, surveys taken in the early 1990s and a referendum on Tatar sovereignty show a strong degree of unity between the two ethnic groups on the question of autonomy.[39]

This fact was not overlooked by Tatarstan's budding political entrepreneurs. During the 1990 elections, candidates often played on growing resentment of Moscow's extraction of Tatarstan's riches (both by Russia and the USSR), and appealed to all ethnic groups to support the sovereignty of Tatarstan. Following the republic elections, the new leaders of Tatarstan tended to follow a civic political strategy which focused on achieving the benefits of sovereignty for the entire population of Tatarstan. While exclusive ethnic nationalists were represented in the new government bodies, their appeal remained restricted to a small sector of society.

Since the early 1990s, there have been strong indications that the dominant definition of the nation in Tatarstan remains a civic one. While the authorities of Tatarstan and Russia have been in almost continuous confrontation over the question of Tatarstan's relationship to Russia, the most recent set of agreements regulating Tatar-Russian relations appears to reflect the continued dominance of an inclusive national identity in Tatarstan.[40] While the new agreements support the demands of Tatarstan's population for greater control over their own economic affairs and wealth, they fall far short of the demands made by the radical ethnic nationalist fringe in Tatarstan. Although it remains to be seen whether these agreements will withstand opposition from Tatar radicals, evidence to date would seem to indicate that civic nationalism remains ascendant in Tatarstan today.

Crimea: *The Russian-Ukrainian-Tatar Knot*

As with Tatarstan and Russia, the enclave of Crimea represented a potential focal point for separatist mobilization within Ukraine. Once again, however, the strength and character of the national identities of the dominant ethnic groups in the region played a central role in shaping the direction of social mobilization in Crimea. Furthermore, as in Tatarstan, instrumental considerations of wealth and economic growth became key factors in determining the extent to which ethnic groups might unify around a common political objective.

The Crimean peninsula represented an exotic and unique region within the Soviet Union. A visit to Crimea meant leaving behind many of the grim realities of Soviet daily existence. Known to most Soviet citizens as a center for tourism, recreation, and relaxation, the peninsula's rugged mountains, lush tropical flora, and spectacular beaches and coastline attracted vacationers from across the USSR. Outside Yalta, an immense pillared monument proclaimed, "Every Soviet citizen has a right to a vacation." The peninsula was dotted with hotels, spas, sanatoria, and the vacation dachas of communist elites and military retirees.

The history of Crimea is an unusual one. While at the time of the breakup of the USSR the peninsula was considered an oblast of Ukraine, this status had been acquired only recently. From the late eighteenth century until 1954, the Crimea had in fact been under Russian jurisdiction. Following Catherine the Great's victory over the Ottomans in 1775, Russia had claimed control over Crimea and consistently asserted its dominance until the Bolshevik Revolution of 1917. After a period of turmoil, the creation of the Crimean Autonomous Soviet Socialist Republic was declared in 1921; the Crimean ASSR was once again designated as a component of Russia. In 1945, following the 1944 deportation of the Crimean Tatars on allegations of collaboration with the Germans, Crimea lost its status as an autonomous republic and was relegated to the relatively low status of oblast. Finally, in 1954, to mark Russia and Ukraine's 300-year-old history of friendship, Khrushchev transferred the peninsula from Russian to Ukrainian jurisdiction.

Thus, as nationalism began to flower across the Soviet Union in the late 1980s, the Crimean population found itself in a confusing situation. What was the national identity of this region? History, language, and ethnic composition might indicate a Russian national identity, and yet for the past three and a half decades the peninsula had been completely integrated into Ukrainian political structures and considered a constituent component of Ukraine. As elsewhere in Ukraine, intermarriage rates between Russians and Ukrainians were high, and due to Russification of the Ukrainian population, distinct ethnic lines between Ukrainians and Russians were difficult to draw. Complicating this situation was the transient nature of much of Crimea's population. Because of its status as a premier resort and retirement area, many people did not live in Crimea on a permanent basis or were relative newcomers to the region. Despite Russia's long-term association with Crimea, most Russians could not claim deep historical and ethnic roots to the land.

A third group could, however, claim a centuries-old attachment to the Crimean peninsula. The Crimean Tatars could trace their roots in the region back to the early thirteenth century and thus had perhaps the most legitimate historical and ethnic claim to the region.[41] Since they had been deported from Crimea in 1944, however, few were left on the peninsula to claim their heritage. It was only with the introduction of perestroika that the Crimean Tatars were finally able to openly protest their deportation and begin returning to Crimea. By the late 1980s, several hundred thousand had returned to Crimea and were demanding land, restitution, and greater political rights in the region.

The complex ethnic history of the region shaped the types of nationalist mobilization that occurred both during the perestroika years and following the breakup of the Soviet Union. This in turn significantly affected the character and development of the region's anti-nuclear power movement.

Anti-nuclear Mobilization on the Crimean Peninsula

Plans to construct a nuclear power station on the Crimean peninsula were already well under way by the time of the Chernobyl disaster. The project was a relatively modest one, with a single VVER reactor planned to go on line between 1987 and 1990 and a second several years later. The station was located in the town of Sholkina at the easternmost point of the peninsula (though it is often referred to as the Kerch AES due to its proximity to the larger city of Kerch). It was hoped that the station would alleviate an energy shortage on the peninsula. In fact, the region produced very little of its own energy and was forced to rely largely on transfers from outside. As of 1991, over a quarter of the Crimea's energy was supplied by nuclear power sent down from Ukraine and Russia. The Crimean nuclear power station (KAES) was designed to enhance the region's energy self-sufficiency.[42]

As in the republic capitals of Lithuania and Ukraine, mobilization in the capital city of Crimea began within the intellectual stratum of society. As in Kiev, parallel mobilizations began to take place among two sectors of the intelligentsia: writers and scientific specialists. Within the writers' community, V. P. Terekhov, a well-known local writer, began to raise the subject of the Crimean AES at Writers' Union meetings in 1987. The discussions on this topic became quite emotional, with a number of writers arguing that Crimea's special identity as a pristine and unique resort area must be protected against the irresponsible policies of Moscow. Interestingly, the writers did not argue for the protection of a people or a nation but rather for the protection of their identity as prime vacation territory. Over time, a small core of concerned writers formed their own discussion circle to further consider the KAES problem.

Simultaneously, a number of scientific specialists began to take interest in the issue. One of the first was solid-state physicist, Anatolii Svidzinsky. Svidzinsky had long been interested in environmental issues and had even had the opportunity to become involved in the work of the Club of Rome in the late 1970s. In 1978, he had presented a paper on global environmental concerns at Simferopol State University which mentioned the problem of nuclear power and waste disposal. Svidzinsky recalls that at the time he had doubts concerning the prevailing doctrine of the "absolute safety" of nuclear power, but the topic was too politically sensitive to refer to in any but the most oblique terms. He searched for data on reactor safety in the late 1970s, but he was unable to find any and consequently dropped his pursuit of this question.

All this changed in 1986, however, with the Chernobyl disaster. In a misguided attempt to prove to the Soviet population that the USSR's nuclear power stations were no worse than those in the West, the Soviet high-circulation press had been flooded with information about nuclear accidents outside the USSR during the months following the Chernobyl accident. Svidzinsky found the information alarming and began to compile this published data and compute statistics on accident rates and probabilities.

In 1987, Svidzinsky shared his findings with other physicists and mathematicians at Simferopol State University. Soon a discussion group was formed which gradually expanded to include scientists from outside the university and outside Simferopol.[43] As the group expanded, geologists, biologists, and a variety of other scientists became involved. Interestingly enough, however, no nuclear physicists ever joined this circle.

By early 1988, discontent with the plans to open a nuclear power station in Crimea began to overflow into the public arena. As elsewhere, diminishing censorship provided the opportunity for anti-nuclear viewpoints to finally start appearing in the local press. Both writers and scientists began to submit their views to the media, and several of the more progressive local papers, including *Slava Sevastopola* and *Krymskii komsomolets,* seemed quite receptive to the opposition platform.

In April, the public was invited to attend an open forum at Simferopol State University at which the KAES issue was to be discussed.[44] The forum was academic in nature, with all of the members of Svidzinsky's discussion group presenting papers on the KAES question. An audience of approximately three hundred turned up for this novel event; several speakers from the forum have categorized the group as primarily fellow academics and intellectuals.

While all of the presentations questioned the wisdom of building a nuclear power station on the Crimean peninsula, few speakers went so far as to suggest a solution to the problem that was rapidly growing in Kerch. Most acknowledged the huge government investment in the station and were reluctant to call for its cancellation so late in the game. Despite the lack of concrete proposals, the forum revealed the strong opposition to the station that had been lurking beneath the surface. Speakers later expressed surprise at the degree of anti-nuclear unanimity in their audience.[45] Unlike in the Lithuanian case, however, this early forum did not represent an opportunity to voice hidden nationalist sentiments. Most speakers and members of the audience stressed the need to preserve Crimea's identity as a vacation mecca, and many referred explicitly to Lenin's famous contention that the Crimea should be preserved as a unique recreational zone.[46]

Following the session, a number of speakers and a handful of members of the audience stayed on to continue the discussions on a less formal basis. The twenty-five people who stayed decided to create a committee to fight for environmental protection of Crimea and particularly to oppose the ever growing KAES. A committee of eight scientists was selected to lead the new movement, including physicists A. V. Svidzinsky, Pivovarov, and A. V. Bruns; geologist E. P. Tikhonenkov; and biologist A. S. Komarov. Pivovarov was elected president of the committee and Svidzinsky's paper laying out the key problems of the KAES was chosen as the preliminary platform for the organization. Identifying themselves as part of an all-union organization, the movement's founders chose to consider themselves a branch of the all-USSR organization Ekologiya i mir (Ecology and Peace).

Interestingly enough, from 1988 to 1989 the question of whether the new organization should be affiliated with Ukrainian or Russian activist networks did not emerge. While activists initially linked their organization to a primarily Russian organization, they were not hesitant to affiliate themselves to a Ukrainian network when the opportunity emerged; when the all-Ukraine organization Zelenii svit was established in Kiev, the Crimean activists also registered their association under the Zelenii svit umbrella. Any connections to the larger world were thought to be of assistance, and during this early period of mobilization, ethnic issues did not appear to play a role.

Following the April forum, anti-nuclear mobilization in Crimea began to pick up pace. Oppositional letters from both writers and scientists began to appear with some regularity in the press. Joint letters from members of the Writers' Union and physical scientists also began to appear, indicating that the two sectors of the intelligentsia were beginning to join forces on this issue.[47] An all-union conference on the economy and tourism held in Yalta, which featured leading Moscow anti-nuclear activist Mikhail Lemeshev, also provided an opportunity for local writers and scientists to become acquainted with each other while simultaneously introducing the new activists to anti-nuclear crusaders from outside the region. This conference was once again given wide publicity in the local media, though reports indicate that the tone of the conference was not as stridently anti-nuclear as the April forum had been.[48]

On May 20, 1988, a second session was held at Simferopol State University to discuss the KAES issue. This time a broader audience responded to the invitation, and the main hall of the university was reportedly packed to capacity with people from all walks of life. Again, local scientists presented serious papers on the hazards of the KAES. This time, however, they also invited the director of the KAES, Tansky, along with nuclear specialists from Moscow to respond to their complaints. Unlike the earlier session, discussions at this public forum became quite heated as members of the audience began to challenge the power industry's representatives.

Following the May forum, the ranks of the Crimean branch of Ekologiya i mir began to expand rapidly. The core of anti-nuclear writers from the Writers' Union quickly joined the growing movement, and scores of concerned citizens from outside the intelligentsia began to set up local chapters of Ekologiya i mir. An article published in *Krymskii komsomolets* (May 28, 1988) noted the explosion of public opposition to the KAES which had occurred in the aftermath of the two forums held at Simferopol State University. The paper claimed that it was being flooded by letters, telegrams, and phone calls and mentioned one letter which came accompanied by over 2,000 signatures. It was also reported that the budding anti-nuclear activists had begun a door-to-door campaign to collect signatures against the station, and it was estimated that over 20,000 people had already signed such petitions. The reporter expressed surprise at this unexpected outburst and its contrast with the previous passivity of Crimean society.

Throughout this entire process of mobilization, the Communist Party played an ambiguous role. The Crimea was known as a conservative communist stronghold. Throughout the perestroika period, regional party leaders strongly resisted Gorbachev's calls to restructure their political, economic, and social realms. Thus, the budding anti-nuclear activists recognized early on that their success would depend on recruiting the party organization to their side. Activists from both the Writers' Union and the scientific community contend that their strategy was from the start a two-pronged one: mobilizing the masses and winning over the obkom of the Communist Party.[49]

Beginning in 1987, both writers and scientists appealed to the Communist Party obkom to assist them in challenging the Crimean AES. Members of the obkom also confirm that the topic of the KAES and its suitability to the Crimean conditions was raised at a number of obkom meetings in 1987 and 1988. While the obkom did not take a stand on this issue until late in the game, many members of the party apparently recognized the growing importance of the nuclear power question and the potential benefits of playing a leading role on this popular issue. The two open forums held at Simferopol State University in April and May of 1988 were in fact sponsored by the Communist Party. The party, however, did not take sides at this point, and the title for both forums was "The Crimean AES: Pros and Cons."

In an effort to resolve the growing tensions surrounding the Crimean AES, V. Kazarin of the party obkom organized a meeting between leading anti-KAES activists and the administration of the nuclear power station on May 26.[50] Even A. L. Lapshin, the USSR deputy minister of atomic energy, made the trip down to Kerch to help win over these troublesome intellectuals. Unfortunately, however, the session was described by many who attended as an old-fashioned propaganda meeting, and activists complained that their views were not heard and their concerns not addressed. Rather than resolving the issue, this session seems to have convinced many of the activists of the futility of attempting to reason with the USSR military-industrial bureaucracy.

Throughout the remainder of 1988, the patterns of mobilization that had been established earlier in the year continued. Intellectuals continued their anti-KAES crusade; awareness and mobilization among the mass public expanded; and the party obkom remained on the fence—a mediator but not yet a player. By the autumn of 1988, however, mobilization in Crimea was beginning to catch the attention of the political elite in Moscow.

News of the growing opposition to the Crimean AES traveled to Moscow by several routes. First, the numerous letters and appeals composed by the scientists and writers to win the party obkom over to their side were usually sent to Moscow as well. Furthermore, many of the petitions that were circulated during the summer and fall of 1988 were sent directly to Gorbachev. Second, word of local opposition to the nuclear power station began to travel upward through existing CPSU structures. In particular, the Nineteenth Extraordinary CPSU Conference which brought delegates from across the entire USSR to Moscow during the summer of 1988 provided an outstanding opportunity for party members to voice local concerns. While little unanimity existed among the Crimean delegation on the KAES issue, at least one delegate, M. Melnikov of the Crimean Agricultural Institute, spoke out against continued construction of the station.[51] In addition, members of the larger Ukrainian delegation appealed to the conference participants to consider a brief one-year moratorium on all nuclear power stations in Ukraine, including the KAES.[52]

Finally, while the anti-nuclear movement on the Crimean peninsula was in fact quite self-contained, this did not prevent it from benefiting from the lobbying and campaigning efforts of the all-Ukraine movement, Zelenii svit. The Crimean AES was officially located within the republic of Ukraine, and thus Kiev activists consistently included the station in their list of demands.

When Moscow reacted to this surge of anti-nuclear sentiment in Crimea, it was only slowly and incrementally. As was generally the case, Moscow's preferred first step was the formation of an expert commission to evaluate the design and safety of the station. Thus, late in the summer of 1988, First Secretary Gorbachev, Chairman of the Council of Ministers Ryzhkov, and Minister of Power Shcherbina, appointed the director of the Kurchatov Institute, Eugenii Velikhov, to head such an investigation. While the commission included numerous Moscow experts and representatives of the power industry, it also brought in a number of local specialists to participate.

Several months later, the commission issued its official report.[53] Its findings largely supported the concerns of local anti-nuclear activists. While a number of shortcomings of the station were noted by the commission, the most serious concern was the possibility of regional seismic activity at levels much higher than the design could withstand. According to a report in *Rabochaya hazeta* (Dec. 23, 1988), fifteen members of the commission voted in favor of cancellation, four members voted for continuation, and two were undecided. Reportedly, the four proponents of continued construction were all affiliated with the Ministry of Atomic Energy.

Despite the commission's findings, however, the USSR Council of Ministers remained reluctant to halt a project so near completion. Despite scientific objections, construction continued. Local scientists complained bitterly that the Ministry of Atomic Energy was indifferent to specialist input, and the Ukrainian Academy of Sciences publicly threw their support in with scientists and called for cancellation of the project.[54] Moscow, however, seemed impervious to scientific objections.

While Moscow seemed prepared to ignore the advice of specialists, they found it more difficult to overlook the explosion of anti-nuclear public opinion that seemed ready to engulf them. In Crimea, the movement against the KAES began to take on a mass character during the winter of 1988–89. Chapters of Ekologiya i mir sprang up in towns and villages across Crimea, and petitions were circulated at an unprecedented rate. Estimates of the number of signatures collected during this period vary according to source, but the figure of 300,000 signatures is often cited. Even taking into account some exaggeration, it seems apparent that mobilization on this issue was occurring at an unprecedented level in Crimea.

During the winter of 1988–89, however, the character of the anti-nuclear movement in Crimea also underwent a dramatic transformation. During most of 1988, opposition to the Crimean AES was led by members of the intelligentsia. Until then, scientists and writers worked together to publicize their technical and scientific concerns about continuing construction on this nuclear power station. The founders of Ekologiya i mir came entirely from the scholarly community, and while they hoped to win the masses over to their cause, their real goal was to use sound scientific arguments to win their case.

By the winter of 1988–89, however, droves of nonintellectuals were joining the movement, and the scientists were quickly outnumbered. All of the scientists and writers who were interviewed describe what happened during this winter in terms of "infiltration" by pensioners. The scientists expressed dismay at the changing character of discussions within the organization. Meetings became dominated by pensioners who had little knowledge or interest in the technical issues. For a short while, the elected leader of the group, Pivovarov, managed to keep control over this new fringe membership. By early 1989, however, he was fed up and quit. Within months, every scientist and writer who had been among the founders of the organization walked away from it in apparent disgust; many claimed that "populism" had replaced reasoned debate.

Interestingly enough, however, with the changing membership of the movement also came a change in tactics. When intellectuals dominated the movement, a variety of tactics were debated and tried. As a rule, the intellectuals were not hesitant to challenge the political authorities. The pensioners, however, were adamantly opposed to any activities which might be viewed as politically challenging. The movement's change from an independent intellectual organization that was willing to challenge the region's political elite to a relatively docile and apolitical organization led to widespread accusations of party infiltration. Such accusations are difficult to substantiate, but it is clear that the movement was channeled in a relatively safe, apolitical, and unchallenging direction from 1989 on. Certainly such an infiltration and redirection of the movement would have been in the party's best interest at the time, and thus the accusations seem plausible (though unsubstantiated). During the spring of 1989, the party also became more supportive of the anti-nuclear cause, and the environment became more hospitable to the new popular movement. Ekologiya i mir's application to register as an informal organization was quickly accepted in early 1989 without any of the red tape that usually accompanies the registration process.

The popular appeal of this issue was once again confirmed during the spring of 1989 with the elections to the USSR Congress of People's Deputies. Because Ekologiya i mir had officially registered, it was permitted the unusual privilege of nominating candidates to the new congress. Six candidates ran in Crimea. The importance of the anti-KAES issue, however, was most vividly demonstrated by the almost complete unanimity in the campaign platforms of all candidates to the USSR Congress of People's Deputies (CPD). Most astute budding politicians recognized the popularity of the anti-KAES platform; few were foolish enough to neglect to include opposition to the station in their electoral platform.

Following the elections to the CPD, party and other official organizations that had been sitting on the fence suddenly swung over to the anti-KAES side. First secretary of the Crimean obkom, Girenko, who had earlier expressed doubts about the feasibility of canceling the KAES, suddenly became a leading anti-nuclear spokesperson. Likewise, the newly formed Crimean branch of the State Committee for the Protection of Nature also came out in favor of canceling the project.

The growing support by the obkom had a dramatic effect on the battle against the Crimean nuclear power station. During the spring of 1989, the local party organization finally threw its full weight behind the anti-nuclear crusaders, and their influence proved substantial. On April 19, 1989, the Crimean oblast soviet, under significant pressure from the party obkom, voted to cancel construction of the KAES and called for its conversion to some other type of industrial facility. In addition to responding to strong party pressure, members of the oblast soviet were undoubtedly also concerned about their reelection prospects for 1990 and responding to the overwhelming popularity of the issue.

Of course, the concrete implications of the oblast soviet's decision were almost nonexistent. Nuclear power decision making was still considered the domain of Moscow, particularly the USSR Council of Ministers, and the local cancellation decision had no impact on the ongoing construction in Kerch. The decision, however, was significant because it demonstrated the strong support of local party and government authorities for the popular anti-KAES cause and forced Moscow decision makers to acknowledge the strength of local opposition to this project. Though a number of local soviets voted to cancel projects in their region in the aftermath of the 1990 local elections, this case in particular was trendsetting in that the oblast soviet took this step a full year before the others, in anticipation of elections.

As the first such local decision in the USSR, the decision by the Crimean oblast soviet to cancel the station undoubtedly took Moscow by surprise. During the summer and fall of 1989, the fate of the KAES was clearly under discussion in Moscow. In late May, it was rumored that a decision had been taken to halt construction of the station. V. Fokin, deputy chairman of the Ukrainian Council of Ministers, noted in an interview that such a decision had been taken.[55] Moscow, however, refused to confirm Fokin's claim, and the question remained unresolved. Several months later, the USSR minister of atomic energy was asked directly about the fate of the KAES in an interview for *Rad. Ukr.* (Radyanska Ukraina) (Aug. 9, 1989) and refused to answer, saying only that the station would not be opened until an international team of experts had deemed it satisfactory.

With construction of the station continuing despite scientific recommendations and local opposition, members of Ekologiya i mir decided to pursue more aggressive tactics. Throughout the summer and well into the fall of 1989, Ekologiya i mir organized picketing of the station. On almost any day for a several-month period, a handful of picketers could be found outside the KAES. In addition, several larger protests were held near the station, the most memorable being a mock funeral for the Crimea as a vacation mecca.

In September, the first secretary of the Crimean party obkom, A. Girenko, sent a formal letter of protest to deputy chairman of the USSR Council of Ministers, Lev Ryabev, demanding to know why construction was continuing in the face of negative scientific evidence.[56] By this time, the party had clearly established itself as a major player in the anti-nuclear campaign.[57] Girenko's public complaint was quickly followed by a second decision from the oblast soviet—this time to halt financing of construction of the KAES.[58] While the decision was primarily symbolic (since most funding flowed from Moscow), it once again passed the message on to Moscow that the local population was strongly committed to preventing the opening of the KAES.

Finally, on October 25, 1989, the USSR Council of Ministers acceded to Crimean demands to cancel the Crimean AES. *Pravda*'s report of the cancellation decision stressed the importance of negative scientific findings combined with strong pressures from Crimean Communist Party Secretary A. Girenko.[59] Due to the fact that the station was nearly complete, the Council of Ministers decreed that it would be converted into a nuclear power training facility and promised that no nuclear fuel would be delivered to the station. The crusade begun by a small core of writers and scientists in Crimea had finally been won.

The National Factor

The anti-nuclear power movement in the Crimea was very short lived, appearing in 1987 and disappearing by the beginning of 1990. Following the decision of the USSR Council of Ministers in the fall of 1989, popular interest in the nuclear power issue fell dramatically. While this is understandable given the apparent resolution of the issue, it is interesting to note that the environmental movement as a whole also withered into irrelevancy at this time. By 1991, only a handful of environmental activists could be found in Crimea and the once vibrant chapters of Ekologiya i mir had all but disappeared. No mass environmental actions were observed in the Crimea after 1989.

The fleeting nature of the anti-nuclear and environmental movements in the Crimea would support the hypothesis that these movements were associated with objectives that went beyond pure environmental concerns. The need for environmental activism did not disappear in 1990, but the movements did. It appears clear that these movements offered a safe outlet for the expression of popular dissatisfaction with government policy. As such, these movements represented the first step along the path to political mobilization. Once society began to participate politically, however, there was no need to focus on "safe" issues; by 1990, it was possible for budding political activists to state their political demands openly and to appeal to the population for support of their political platforms.

As was observed in the preceding studies of Lithuania, Armenia, and Ukraine, the tendency to use anti-nuclear and environmental protest as a surrogate for other forbidden demands was often associated with the emergence of nationalism in the non-Russian regions of the USSR. The question thus arises, in this predominantly Russian region of Ukraine, was anti-nuclear and environmental activism also a front for hidden nationalist aspirations? Interestingly enough, there is very little evidence to support claims linking anti-nuclear activism and nationalism in the Crimea. In fact, the anti-nuclear movement in the Crimea appears to have been remarkably free of ethnic nationalist orientation.

Throughout the three-year struggle against the Crimean AES, ethnic Russians and Ukrainians fought side by side to eliminate the perceived threat to the peninsula's identity as a premier resort and recreation area. As was noted earlier, environmental activists showed little concern for the question of ethnic affiliation, formally linking their organizations to both the predominantly Russian association, Ekologiya i mir, and the all-Ukraine group, Zelenii svit. Only in 1990 did the environmental movement split into two separate groups: Ekologiya i mir and Zelenii svit. This decision to split, however, seems to have had far less to do with ethnic identities than with a simple personality clash among the leadership. Sergei Shuvainikov, one of the most active opponents of the Crimean AES, decided to break away from the mainstream of the movement after a falling out with other leaders. He thus declared himself the leader of Zelenii svit and proclaimed the separation of Zelenii svit from Ekologiya i mir. Apparently, however, no other members followed Shuvainikov: the entire membership retained its original organizational structure while a single individual suddenly claimed leadership of a separate Zelenii svit organization. Thus, while on the surface it might appear that the Crimean environmental movement split into Russian and Ukrainian factions in 1990, the reality does not support this hypothesis. It is also interesting to note that Shuvainikov attributed the split to ethnic causes during a 1991 interview, noting that his preference for affiliation to Zelenii svit was linked to the fact that this was an all-Ukraine organization rather than a predominantly Russian association emanating from Moscow. This explanation holds little credibility, however, when judged by later events. During the spring of 1994, Sergei Shuvainikov ran for president of Crimea on a rabid Russian-nationalist platform, earning him the nickname the "Crimean Zhirinovsky!" Thus his 1991 claim that his departure from the mainstream of the environment movement was linked to his preference for a Ukrainian affiliation holds little water!

The only ethnic distinction visible in the anti-nuclear and environmental movements in Crimea was the division between the dominant Slavs (Russians and Ukrainians) and the minority of Crimean Tatars. The Crimean Tatars, who began to return to the region from their Central Asian exile as early as 1986, had a completely different political agenda than the Russians and Ukrainians. The Crimean Tatars were concerned with resettlement issues—housing, land, government subsidies—and had little interest in peripheral questions such as nuclear power and environmentalism. While Crimean Tatar organizations were repeatedly invited to participate in anti-nuclear activities, they consistently declined.

The Crimean Tatars' refusal to hide behind a surrogate issue and commitment to pursue their goals openly and forcefully was linked to several factors. The Crimean Tatars, in fact, had a long history of open confrontation with the Moscow authorities. While in exile in Central Asia, the Crimean Tatars had continuously worked to maintain their separate ethnic identity and to openly pursue their goal of returning to their homeland in Crimea. During the 1960s and 1970s, despite adverse political conditions, a strong association of Crimean Tatars emerged to fight for protection of the Tatars. Leading members of the organization were viewed as dangerous dissidents by Moscow and were frequently imprisoned. This tradition of direct confrontation and dissidence carried on into the perestroika period. The Crimean Tatars were in fact one of the first groups to openly demonstrate outside the Kremlin walls during the first years of perestroika. As they returned to Crimea, they saw no need to change to the safer tactics of surrogacy and instead continued their policy of openly campaigning for greater rights for their people.

Within the Crimean population, a wide gulf separated the dominant

Slavs from the Crimean Tatars. This gulf was fed by the Tatars' own insistence on maintaining separate cultural, linguistic, and ethnic identity. Since the beginning of the return of the Crimean Tatars during the early perestroika period, Tatar organizations had continuously maintained their commitment to supporting the Tatars' distinct ethnic identity, fighting for the assistance in resettlement that they believe was owed to them, and ultimately, for the creation of a "Tatar state" in Crimea. The Crimean Tatars viewed the Russians and Ukrainians in the region as recent émigrés into their territory and demanded that the region be returned to Tatar control.[60]

Although the sharp ethnic split between the Crimean Tatars and the Slavic population of Crimea is cause for concern, it must be noted that the Tatars constitute only a small percentage of the peninsula's population.[61] The real cause for concern lies with the potential for the growing schism within the Slavic population itself—that is, the possibility of growing antagonisms between the ethnic Russians and Ukrainians. This study, however, showed little sign of a deep Russian-Ukrainian ethnic divide during the perestroika period. As noted above, the environmental movement was not divided along ethnic lines (as opposed to what was observed in Lithuania and Armenia). Furthermore, the anti-nuclear debate contained none of the ethnic rhetoric observed elsewhere. Moscow's decision to build a nuclear power station in Crimea was never referred to as a policy of "genocide," and neither Russians nor Ukrainians were particularly targeted as the perpetrators of this evil act. Throughout the debate on the Crimean AES, activists continuously stressed the need to protect the region as a pristine, recreational zone, and resisted any tendency to link anti-nuclear demands with Russian or Ukrainian ethnic identities. In addition, outside the anti-nuclear movement, very little ethnic mobilization of Russians or Ukrainians was observed during the perestroika period. While the Crimean Communist Party set up a series of cultural "clubs" (Russian, Ukrainian, Tatar, and others) in 1991 in order to co-opt any nationalist strivings that might emerge, there was little public response to this initiative, and with the exception of the region's Crimean Tatars, ethnic mobilization on the peninsula remained a tiny, fringe phenomenon.

Rather than acting as a surrogate for nationalism, the anti-nuclear movement in Crimea represented the first step toward political mobilization of the population. While reflecting the goals of greater self-determination and decision-making rights for the region, the anti-nuclear movement exhibited no ethnic nationalist orientation. Following the mass mobilizations against nuclear power in Crimea, however, the population was quick to move on to more political objectives. Because the region was viewed as a magnet for both Soviet and foreign tourists, early political goals in the region focused on giving the peninsula more control over its own affairs and, particularly, the right to reap the profits of a potentially immense tourist industry. The first step along this path was to regain the region's status as an autonomous republic rather than an oblast of Ukraine. This status had been taken away in 1944–45 with the deportation of the Crimean Tatars from the region.

In January 1991, a referendum was held in Crimea asking whether the population supported renewing the region's status as an "autonomous soviet socialist republic" of the USSR. Interestingly enough, this referendum reflected not only demands for enhanced status and decision making in the region but also fears of the growing independence movement in Ukraine. By early 1991, it was becoming obvious that aspirations for independent statehood in Ukraine were on the rise and might eventually lead Ukraine to break away from the USSR. In Crimea, however, the prospect of being forced to join Ukraine in leaving the USSR filled the population with great trepidation. First, only a minority of the population was ethnic Ukrainian and an even smaller minority described themselves as Ukrainian speakers. As Ukrainian nationalism in Ukraine grew, the population of Crimea began to fear the imposition of an alien culture and language in the region. The passage of the Ukrainian language law in 1989 did little to calm fears of Ukrainian national imperialism in the region.

A second factor which is thought to have attributed to the growth of resistance to membership in an independent Ukraine was the conservative character of Crimea's communist leadership. Seeing the reform movement grow around them, the communist leadership of Crimea hoped to shelter themselves from its onslaught by maintaining the political autonomy of the region. Thus, the Crimean Communist Party leadership strongly supported and lobbied for a referendum which would insulate the region from the forces of perestroika and allow the Crimea to participate as an equal in the Union Treaty that was then under debate. The referendum was highly successful, with 93 percent of those who participated voting to upgrade the Crimea's status to that of an ASSR within the USSR and to support Crimea's inclusion as an equal member in the Union Treaty.

Following the referendum on Crimea's status, apprehension about the fate of the region in the event of Ukraine's secession from the USSR grew rapidly. It was only at this time that ethnic factors began to intrude on Crimean politics. Suddenly, the distinction between ethnic Russians and Ukrainians began to receive attention. It was not so much ethnicity that divided the population, however, as language. Less than 4 percent of Crimea's population considered themselves to be Ukrainian speakers,[62] and the institution of Ukrainian as the republic's official language seemed to demonstrate that the Crimean population, as a small percentage of the total population of Ukraine, was likely to find itself a helpless victim of the growing Ukrainian nationalist sentiment in Kiev.

Despite the growing hostility to affiliation with an independent Ukraine, however, ethnic nationalism was slow to emerge in the region. Contrary to trends observed elsewhere in the USSR, neither a strong Russian nor Ukrainian nationalist movement emerged in Crimea during the perestroika period. There was no spontaneous explosion of nationalist sentiment as observed in the Baltics and much of the Transcaucasus. Instead, nationalist sentiment in Crimea grew only slowly in response to the growing assertiveness of Ukrainian nationalists outside Crimea.

Following the breakup of the USSR in late 1991, resistance to inclusion in Ukraine has grown on the Crimea peninsula. The absence of a strong ethnic movement during the perestroika period, however, would support the hypothesis that anti-Ukrainian sentiment in Crimea is based not on a deep ethnic divide within Crimean society but rather a pragmatic assessment of the pros and cons of affiliation with the newly independent Ukraine. On a purely practical level, much of the Crimean population can see little benefit of incorporation within Ukraine. Their language and culture may be under threat from Ukrainian nationalists outside Crimea, their region has little political voice in Ukrainian politics, and the Crimean peninsula lacks total control over its economic resources. Furthermore, with most of the Crimean population having emigrated from Russia, the new trade and travel barriers between Russia and Ukraine have made the simple tasks of telephoning family, mailing gifts, and visiting friends and relatives in the now foreign country of Russia both difficult and expensive. Finally, Ukraine's deteriorating economy and its failure to take any steps that might put the country on the road to economic recovery and prosperity have greatly increased Crimean opposition to affiliation with Ukraine. On a purely practical level, many in Crimea argue that economic hardships in the region would be greatly alleviated through either independence or affiliation with Russia.

While opposition to inclusion within Ukraine has clearly grown in the several years since the dissolution of the USSR, this resistance should not be taken as a sign of growing ethnic distance between Crimea's Russian and Ukrainian population. Interestingly enough, both ethnic Russians and Ukrainians support seceding from Ukraine, and election results in 1994 demonstrated no ethnic orientation in the voting patterns of the Russian and Ukrainian populations.[62] Furthermore, the surge of pro-Russian ethnonationalism that was displayed by Crimean president Yurii Meshkov and the Crimean parliament after the 1994 elections proved unexpectedly short-lived; by June of 1995, these political forces were clearly in retreat and the parliament had largely conceded defeat in its attempt to redefine Crimean politics along pro-Russian ethnic lines. The mass appeal of Russian ethnonationalism was clearly much weaker in Crimean society than many politicians had previously suspected. Thus, the absence of Russian and Ukrainian ethnic mobilization in the struggle against the Crimean AES has proven to be a solid indicator of the depth of ethnic division within Crimean society.

Conclusions: *Anti-Nuclear Mobilization in the National Enclaves*

In comparing the mobilizational experiences of Tatarstan and Crimea, only minor differences in the patterns of social activation were observed. The primary distinction between the paths followed in these two enclaves lay in the differing linkage between nationalist mobilization and the anti-nuclear movement. In Tatarstan, a weak but observable linkage was noted between the growth of a sense of Tatar national identity and anti-nuclear mobilization. The anti-nuclear movement appeared to play a significant role in reviving a distinctive sense of Tatar culture and identity.

In contrast, the anti-nuclear movement in Crimea displayed no linkage to the growth of any national movement in the region, whether Ukrainian, Russian, or Crimean Tatar.

Despite this variation in the linkage between nationalism and anti-nuclear mobilization in the two enclaves, however, the mobilizational patterns displayed in the two enclaves were remarkably similar. While the rebirth of a distinctive sense of the Tatar national identity was observed in Tatarstan, the anti-nuclear movement was only weakly divided along ethnic lines. In fact, both Russians and Tatars participated side by side in the crusade against the Tatar AES, and the dominant trend in the Tatar national movement was based on a civic, inclusive definition of the national group. Russians were neither treated as the perpetrators of nuclear genocide nor excluded from the mobilizational process (in contrast to the pattern observed in Lithuania). Likewise, in Crimea, the anti-nuclear movement was not cleaved along ethnic lines, and both Ukrainians and Russians joined forces to fight Moscow's dominance over the Crimean peninsula.

In both Tatarstan and Crimea, instrumental calculations clearly took precedence over ethnic exclusivity. The drive for autonomy and independence was based not on the desire for the creation of an ethnically based nation-state but rather on perceptions of the economic benefits of independence. In both enclaves, the dominant national groups were unified in their goal to break free from Moscow's domination and claim control of local economic resources for the republic's population. In Tatarstan, Tatars and Russians alike dreamed of an independent Tatarstan in which they controlled and reaped the profits of the region's vast oil resources and extensive industrial capacity. Likewise, on the Crimean peninsula, Russians and Ukrainians were united in their goal of controlling the immense tourist potential of this vacation mecca. Only a minority group of Crimean Tatars were committed to the creation of an ethnically based independent nation-state on the peninsula.

The tendency to define the nation in civic rather than ethnic terms that was observed in these two enclaves may be traced to several factors. In Crimea, the lack of clear and widely accepted boundaries between Ukrainian and Russian national identities, observed elsewhere in Ukraine, played a key role in preventing the cleavage of Crimean society along ethnic lines. Thus, as was the case on the Ukrainian mainland, ethnic identities provided only a weak basis for mobilization. While the Tatars of Tatarstan did not suffer from the ethnic blurring observed in Ukraine, they nonetheless found themselves poorly prepared to mobilize along ethnic lines in the 1980s. Although Tatar language, culture, religion, and history were certainly distinguishable from their Russian counterparts, the Tatar identity had been buried under so many centuries of Russification that it was difficult to unearth and revive. The younger generation knew very little of their Tatar heritage and high levels of intermarriage prevented ethnic exclusivity from dominating social interactions. Tatarstan's status as an autonomous soviet socialist republic rather than a full union republic had ensured that the indigenous national group would receive fewer privileges and entitlements than indigenous nationalities of union republics. During the preceding several decades, few opportunities were available for maintaining the national language and culture, and promoting the growth of a powerful and privileged national elite. Thus, the Tatars found themselves less prepared to mobilize on the basis of a unified sense of ethnic national identity than many of their counterparts in the union republics.

Since this study includes only two of the numerous national enclaves of the former USSR, caution must be used in extrapolating to the dozens of minority nationality groups across the fifteen Soviet successor states. I would suggest, however, that the absence of opportunities for the creation of a privileged national elite and for the maintenance of local language and culture in the enclaves that lack union republic status would imply less preparedness to mobilize rapidly along ethnic lines. In some cases this may simply mean a delay in ethnic nationalist mobilization, while in others there may be significant hope for the victory of a civic sense of national identity and mobilizational patterns based on instrumental calculations of benefit rather than ethnic exclusivity. As always, however, differences in national identity, culture, and history may lead to variations in expected trends, as recent events in Chechnya testify.

CONCLUSIONS

The convergence of environmentalism and nationalism in many regions of the former USSR created the potential for the emergence of powerful mass movements during the perestroika period. Because environmental activists could appeal not only to popularly held environmental values but also to people's sense of national identity and community, the environmental movements of the perestroika period attracted far broader constituencies than would otherwise have been the case. Rather than approaching the population in terms of abstract environmental ideals, activists were able to present the nuclear power issue as a very real and material threat to the survival of a specific territory and group of people. In cases in which national identity was highly consensual, it proved quite simple to portray nuclear power as an imperialist threat to national survival.

The emergence of the eco-nationalist phenomenon in the former USSR was an outgrowth of both the unusual structural conditions of the perestroika period and the strength of the national identities that had survived the Soviet period in some of the republics and regions of the USSR. Were these conditions unique however? Was eco-nationalism simply a product of the unstable and unusual features of the late-communist period in the Soviet Union, or might we expect to see it elsewhere around the globe? In the case of the Soviet anti-nuclear movement, the power of the movement emerged from the fact that the nuclear threat could be easily translated into a symbol of the domination of one ethnic or political group over another. The poorly constructed and operated nuclear power stations were obvious symbols of Moscow's disregard for the welfare of its member nations. They represented the unequal relationship between Russians and other ethnic groups, and between the central Soviet authorities and republic and regional leaders. In this case, national inequalities and environmental complaints became synonymous.

This situation is not unique. In numerous inter-state environmental battles as well as domestic struggles between regions or ethnically defined territories, the potential certainly exists for environmental struggles to take on nationalist overtones. Whether or not these two causes will converge, however, depends on many factors. As in the Soviet case, structural conditions which might force nationalists to mask their intentions behind a surrogate cause might favor the convergence of environmentalism and nationalism. Likewise, communities which hold a strong, consensual sense of national identity and distinctiveness would be more likely to introduce nationalism into the environmental crusade. In addition, a history of hostility or inequality between national or regional groups involved in the environmental contest would also add to the potential for eco-nationalism. The most unpredictable factor, however, lies in the role of individuals. As we have observed across Eastern Europe and the former USSR in recent years, the emergence of effective political entrepreneurs willing to play the nationalist card is often a critical feature in determining the extent of nationalist mobilization in a society.

The convergence of environmentalism and nationalism may have both positive and negative implications for the success of the environmental crusade in a given region. On the positive side, by appealing to a sense of national identity and community, the environmental movement may greatly expand its constituency and thus its influence. Unfortunately, however, the negative implications of this phenomenon are likely to far outweigh the positive. First, by adding a very tangible environmental threat to already existing national tensions, exacerbation of nationalist antagonisms is likely. In the Soviet case, the convergence of environmentalism and nationalism did not lead simply to the victory of the environmental movement; it also lead to the nationalist fragmentation of the USSR. While many may agree that in the case of the USSR, the breakup of the empire was not a tragedy, this would not necessarily be the case with other ethnic or regional conflicts around the world.

In addition, the eco-nationalist phenomenon implies a certain superficiality in popularly held environmental values. As we saw in the Soviet case, when the appeal to a sense of national community was eliminated, interest in environmental issues plummeted. Thus eco-nationalism can be deceptive, leading one to believe that a high degree of environmental consciousness exists, where in fact there is little or none. In the Soviet case, the aftermath of the heady days of eco-nationalist (and eco-regionalist) fervor has left disappointment and disillusionment in its wake.

In reviewing the cases presented in this study, however, it should be noted that the extent to which anti-nuclear activism and nationalism converged in Lithuania, Armenia, Ukraine, Russia, Tatarstan, and Crimea varied considerably across republic and region. Thus, let us turn now to this variation. In the following section, I will briefly evaluate the extent to which the theoretical framework presented in chapter 1 accounted for the observed characteristics of the anti-nuclear movements studied, then consider what lessons may be drawn from the variation in the linkage between anti-nuclear activism and nationalism that was observed in these cases.

Structure and Identity Revisited

In the preceding chapters, strong similarities in social movement characteristics and developmental patterns were observed across republics and regions of the former Soviet Union. These similarities tend to support the hypothesis that structural factors have played a very significant role in shaping patterns of social activism in these late- and postcommunist societies. Because all of these regions experienced similar resource availability and distribution during the perestroika period and only slightly diverging resource and opportunity structures following the breakup of the USSR in 1991, the observed similarities lend credence to the resource mobilizational approach suggested in chapter 1.

In reviewing the preceding cases, it is immediately obvious that all of the indigenous movements displayed substantial organizational shortcomings during the perestroika period and after. This supports the suggestion, made earlier, that widespread impoverishment of these late- and postcommunist societies, accompanied by the low access to key tangible and intangible resources of new independent actors relative to established Communist Party elites, would likely result in classical social movement organizations based on voluntary leadership, with very weak organizational linkages both within and between chapters. As was observed in these studies, difficulties in obtaining access to the media and independent printing facilities, lack of adequate meeting space, and inadequate communications equipment severely limited the ability of new social movements to organize efficiently and effectively. Even more than the missing tangible resources, however, these movements clearly suffered from the activists' lack of independent mobilizational experience. Rather than having large preexisting networks to mobilize quickly to their cause, these new movements were lucky if they had a tiny network of concerned intellectuals to provide the initial core of the movement. In most cases, these amateur movement entrepreneurs also lacked organizational skills. During the perestroika period, the anti-nuclear movements were dominated by concerned citizens whose experience in organizing mass movements was practically nonexistent.

The organizational shortcomings observed during the perestroika period largely carried over to the post-1991 transition period. While the Communist Party's iron grip on mobilizational resources was largely eliminated by 1991, independent actors still found it difficult to obtain the facilities and equipment necessary for effective mobilization. As the GNP continued to contract across most of the former USSR, resource availability remained highly limited for many indigenous groups. The most interesting aspect for the post-1991 period, however, was the changing leadership of the anti-nuclear and environmental movements of Moscow and Kiev. As the newly independent states began to remove restrictions on foreign economic activities within their territory, chapters of preexisting international environmental organizations began to pop up to fill the vacuum left by the decline of the indigenous movements. The most notable example has been the expansion of the international group Greenpeace in Russia and Ukraine. While small chapters of Greenpeace existed in Moscow and Kiev prior to 1991, these organizations have become the most dynamic and active environmental social forces on the horizon today. With almost-total foreign sponsorship, these groups have access to far greater mobilizational resources than their indigenous competitors. Substantial inflow of funds, photocopying and fax machines, computers, and organizational expertise have created organizations quite unlike any before seen in Russia and Ukraine. To date, however, these organizations are largely restricted to the capital cities, and they have not had a major impact on environmental organization in less cosmopolitan regions. A recent inflow of foreign capital to support struggling indigenous environmental groups also has the potential to alter mobilizational patterns in the newly independent states. Whether or not these outside funders will significantly shape agendas, tactics, and organizational characteristics of indigenous groups has yet to be seen. What is clear is that since 1991 foreign capital has become the single most important determinant of group survival and success.

The changing tactics observed in these regional case studies also reflected changes in both resource availability and opportunities. In both Lithuania and Armenia, preferential access to key resources for movements deemed nonthreatening by the Communist Party combined with a strong sense of national identity in these republics, led to the phenomenon of movement surrogacy during the perestroika period. Because the anti-nuclear platform could be used to mobilize people to defend their nation against the potentially genocidal policies of the imperial center, the anti-nuclear and nationalist causes tended to meld together in both Lithuania and Armenia. Movement surrogacy, however, was a short-lived phenomenon; as resources shifted from environmental to more radical political platforms the need to utilize a surrogate tactic disappeared. Thus, in Armenia the anti-nuclear movement lasted only a few weeks, while in Lithuania it survived little more than a year; in both cases, activism shifted in the direction of openly nationalist mobilization as soon as the political leadership demonstrated their unwillingness to crack down on overt nationalism. Interestingly enough, surrogacy did not emerge as an important tactic in Ukraine, Russia, or the national enclaves of Tatarstan and Crimea. In these cases, the absence of a strong and consensual ethnic national identity impeded the ability of movement entrepreneurs to utilize the anti-nuclear movement for purely nationalist objectives. The way in which the anti-nuclear and nationalist movements of a region were linked tells us a great deal about the strength and character of national identity in that area — a subject we will return to below.

In addition to the temporary nature of the surrogacy tactic, other methods used by the anti-nuclear movements during the perestroika period and after also changed over time. Initially, the extreme lack of resources for independent actors combined with a dearth of opportunities for influencing political decisions led to the predominance of disruptive tactics. Mass protests, commonly known as "meetings," were the order of the day during the early perestroika period. Thousands of people took to the streets to demand government actions on a number of fronts, including nuclear power. With the elections of 1989 and 1990, however, new opportunities for influence emerged. Electoral politics became a major forum for the anti-nuclear crusade. In the all-union elections of 1989 and, particularly, the republic and local elections of 1990, anti-nuclear platforms were almost unanimously adopted by competing candidates. The republic and local elections of 1990 were rapidly followed by a slew of decisions by republic, oblast, and city soviets declaring their intention to halt nuclear expansion programs in their regions. Thus the electoral strategy was demonstrated to be highly effective.

Following the 1990 elections, another change in tactics that was observed was a shift from mass protests to internal lobbying and direct participation in nuclear decision making. Due to the more open democratic institutions which began to function locally in 1990, anti-nuclear and environmental activists often found ample opportunity to establish themselves as advisors to both individuals and political bodies. This made specialist and intellectual activists a far more valuable commodity and relegated the less-educated mass participants to the sidelines. For many activists, mass protests were no longer viewed as the most effective method for influencing decisions. This was an important factor in the decline of mass activism in 1990 and beyond. In addition, many of the movements lost their most dynamic members as a result of the 1989 and 1990 elections; as their leaders were elected to these newly democratic institutions, the movements found themselves abandoned and leaderless. This again supported a shift in tactics away from mass activism.

Finally, in the aftermath of the breakup of the USSR the shift from mass activism to more sophisticated lobbying techniques has been quite pronounced. Not only have opportunities for lobbying and influence expanded with the growing consolidation of democratic institutions and practices across much of the former USSR but the influx of foreign sponsored organizations has led to the introduction of highly sophisticated tactics already tested outside the region. Not only has Greenpeace engaged in extensive lobbying efforts in Moscow and Kiev but they have also attempted to duplicate the German experience and utilize the courts to halt nuclear expansion. The decision by the Russian government to overturn the parliament's five-year moratorium on nuclear construction and to reinitiate the nuclear power program in late 1993 has been challenged by Greenpeace in the courts system. As yet, however, these legal tactics have yielded few concrete results.

While the resource mobilization perspective clearly yields a great deal of insight into the characteristics and development of the anti-nuclear power movements of the former USSR, the picture is nonetheless incomplete. While strong similarities in movements were observed across regions, the differences in the linkage between anti-nuclear activism and nationalism cannot be overlooked. These divergences in mobilizational patterns bring us back to the question of the role of identity and its contestation in shaping how movements evolve.

In the preceding case studies, we observed a broad spectrum of relationships between anti-nuclear activism and nationalism. At one end of the spectrum, the anti-nuclear movements of Lithuania and Armenia appeared to be little more than temporary fronts for forbidden nationalist demands. At the other extreme, in Russian regions of the Russian Federation and on the Crimean peninsula, the anti-nuclear movements had little or no connection to ethnic nationalism. Between these two extremes, the anti-nuclear movements of Ukraine and Tatarstan displayed a certain affinity for nationalist demands but nonetheless maintained their own distinct identity. This variation in the way nationalism entered the anti-nuclear movements illuminates the differences in the strength and character of national identity in the republics and regions of the former USSR.

As was noted in chapter 1, social movements may be viewed as forums within which people can explore, contest, and reaffirm identities. In joining the anti-nuclear movement, one of the first social movements to emerge after decades of oppression, people clearly were concerned with more than just nuclear safety. While Chernobyl and the threat of a nuclear disaster may have provided the initial impetus for mobilization, the movement eventually became a way in which participants came to understand their new identity in a rapidly changing world. As the old order was first challenged then rapidly discredited during the perestroika period and after, people were left disoriented. With their cognitive maps shattered, citizens of the former USSR were forced to take a new look at their world and reassess their own identity within it.

In mobilizing against Moscow's nuclear power decisions, movement participants had the opportunity to explore not only their attitudes toward the environment but also their own roles as members of a political community. The issues that they implicitly questioned were: what is the nature of our political community? who are its citizens? what is the appropriate relationship between citizens and the state? While the answers to the first two questions varied, participant members were largely in agreement on the third; in participating in this kind of unprecedented independent activism, people were confirming their belief that citizens should have a much greater role in political decision making. While contesting and exploring precisely how far the shift in power from state to citizenry should go, members agreed that the state's monopoly over decision making should be eliminated. In all of the cases observed, resentment of Moscow's complete dominance over local decision making played a key and explicit role in mobilizing opposition to nuclear power.

While movement participants largely agreed on the need to shift power from the state to society, it was often much more difficult for them to reach a consensus on the nature of political community and its membership. What political community did they identify with? The USSR? Their republic? Perhaps their territory or oblast? And who were the members of their political community? Was membership limited to a particular ethnic group or open to all residing in the region? In essence, they were asking, who is "us"? These questions of political community and membership were central to anti-nuclear activists in all of the cases observed.

In Armenia and Lithuania, movement participants moved toward a consensus on these questions much more quickly than in the other republics and regions studied. In both cases, activists quickly agreed that the appropriate political community was the republic rather than the union. Furthermore, membership in the community was largely assumed to flow from ethnic factors. In Armenia, the short-lived anti-nuclear movement was rapidly transformed and incorporated into the strongly nationalist Nagorno-Karabakh Committee. This very popular committee celebrated Armenia's distinctive history and culture and demanded the reunification of the Armenian people. The demands for republic sovereignty which emerged almost immediately and an overwhelming emphasis on ethnicity in campaigning for a return of the Nagorno-Karabakh region demonstrated the high level of social consensus on the appropriate nature and membership of the political community in Armenia.

Similarly, in Lithuania, it was not long before it became obvious that members of the anti-nuclear movement considered themselves citizens of Lithuania, not the USSR. From its very inception, the movement portrayed the nuclear power issue as an example of Moscow's imperial treatment of Lithuania. People were called upon to defend the Lithuanian land and people against the destructive policies of the center. Even before the anti-nuclear movement was swallowed by the nationalist Sajudis organization, the movement's platform, as espoused in open forums and the media, reflected an overwhelming emphasis on issues of Lithuanian nationhood and sovereignty.

The anti-nuclear movement of Lithuania also offered people the opportunity to explore the question of who should be considered citizens of Lithuania. As the movement grew, and particularly after it was absorbed by Sajudis, the anti-Russian orientation became ever more prominent. Because the Ignalina AES was built and staffed primarily by Russians, the nuclear power issue provided a powerful focus for anti-Russian hostility. The area around the station was repeatedly referred to as an example of "demographic pollution," and anti-nuclear articles often highlighted the Russian role in promoting these "genocidal" nuclear power policies in Lithuania. While ethnicity clearly played a determining role in defining Lithuania's national community for some, however, there is also evidence that other movement participants were reluctant to exclude Russians and other ethnic groups from the citizenry. Thus, while the anti-nuclear movement seemed to be moving in the direction of an exclusive ethnic identity for Lithuania's political community, this question remained contested.

In addition to providing a forum for core members to explore identity, the anti-nuclear movement played an important role in opening this debate to the wider society. While intellectual elites who founded both the anti-nuclear and national movements of Lithuania clearly began with a strong sense of their distinctive Lithuanian identity, the bulk of the population needed to be prodded into reconsidering who they were. After almost fifty years of Soviet domination, much of society appeared dormant. The anti-nuclear crusade offered a perfect tool to mobilize society and to cultivate a revived sense of national identity in the population. By beginning with a safe, apparently apolitical, topic, activists were able to appeal to large sectors of society who might not have been ready to risk involvement in a more radical movement. In addition, by using the nuclear power issue as a way to graphically demonstrate the potential dangers of Soviet domination, activists were able to involve people who might not have initially favored an independence platform. The anti-nuclear front allowed nationalist activists to move gradually in reawakening society and pulling them over to their cause.

The rapid transformation of the anti-nuclear movements of Armenia and Lithuania into national sovereignty movements indicated a strong degree of consensus among their leaders on issues of national identity. Intellectual elites in both republics already possessed a clear sense of who they were; as soon as opportunities permitted, they were quick to discard the anti-nuclear cause and move on to their true demands. There was little need for a protracted debate as to whether they were citizens of the USSR or their republic—their answer was obvious. The unusual consensus on national identity observed in these two republics contrasts sharply with the other cases examined here.

As we turn to Ukraine and the national enclaves of Tatarstan and Crimea, the contrast is immediately apparent. Whereas intellectual elites of Armenia and Lithuania possessed a clear sense of national identity early on and acted as movement entrepreneurs in mobilizing society to the nationalist platform, this was not the case in Ukraine or the national enclaves. In these cases, both intellectual elites and the mass of society found themselves confused about their primary political identification. As independence movements began to shake the Baltics, intellectuals and nonintellectuals alike began to question whether perhaps their republics or regions should also be independent political entities. The debate, however, was relatively slow to surface and remained highly controversial all the way up until 1991 (in the case of Ukraine) and to the present for both Tatarstan and Crimea.

In Ukraine, anti-nuclear activists were reluctant to tie their demands too closely to Ukrainian nationalism. While the opponents of nuclear power referred frequently to Moscow's colonial treatment of Ukraine, activists were also quick to point out that they were not advocating Ukrainian independence. Even in early 1991, the leadership of Zelenii svit explicitly dissociated itself from more radical calls for Ukrainian secession.

Interestingly enough, however, there is evidence that this issue was hotly contested. The constant organizational crises and leadership scandals that plagued the movement in 1990 and 1991 were often explicitly linked to disagreement about the extent to which Zelenii svit should identify with both Rukh and radical calls for national independence. Thus the anti-nuclear movement did provide a forum through which people might explore issues of national identity. And, as in Lithuania, it provided an effective tool for reawakening a long-dormant society and initiating discussion on Ukraine's proper relationship to Moscow. Unlike the Lithuanian case, however, a consensus was not soon reached. Even as independence was achieved in December 1991, substantial evidence indicates that much of society felt ambivalent or confused about Ukraine's new political status.

In searching for a sense of identity, anti-nuclear activists in Ukraine also had to confront the ethnic dilemma; in a republic with such close historical and cultural ties to Russia and more than a quarter of the population considered ethnically Russian, how should the national community be defined? On this question, anti-nuclear activists tended to shy away from an ethnic definition of the Ukrainian nation. The anti-nuclear protest movements in fact tended to incorporate both the Ukrainian and Russian populations of the areas surrounding the nuclear power stations. Unlike in the Lithuanian case, nuclear power stations were not portrayed as demographic abominations. Anti-Russian rhetoric did not form a significant component of the anti-nuclear debate in Ukraine. Even in the movements opposing stations in West Ukraine (e.g., Rovno AES), where ethnic homogeneity and nationalism were thought to be stronger, little sign of ethnic exclusivity was observed among anti-nuclear activists.

The evolution of the anti-nuclear movement of Tatarstan followed a very similar pattern. As in Ukraine, the movement provided a forum to contest the nature of the Tatar identity. Where was Tatarstan to fit in the new order? Was it to be a component of the USSR? Of Russia? Or an independent political entity? Should membership in the Tatar nation that appeared to be emerging in the late 1980s and beyond, be limited to ethnic Tatars or also extended to the immense Russian population of Tatarstan? While the confusing developmental path of the anti-nuclear movement indicates that these issues were not easily or immediately resolvable for the Tatar population, the resistance of the population to attempts to mobilize ethnic exclusivity were encouraging. As we look at Tatarstan today, it is clear that these questions are still being contested; Tatarstan's status within Russia has yet to be accepted by all members of the republic's population.

Similarly, in the Crimea questions of national community and identity have yet to be resolved. As in the Tatar case, the anti-nuclear movement provided a useful medium for exploring Crimea's place in the USSR and what it meant to be "Crimean." Due to Crimea's unusual history, however, activists found themselves at a loss to answer these questions. While all soon agreed that Crimea should have greater economic autonomy, the question of Crimea's proper place in the Commonwealth of Independent States has yet to be answered. While the anti-nuclear movement has long since disappeared from the peninsula, the issues of identity that it raised are still very hotly contested. By early 1995, Crimean Tatars were demanding a Tatar state, while Russians in Sevastopol had declared themselves part of Russia, and parliamentarians in Simferopol were openly defying Ukrainian control. While ethnic nationalism has thus far failed to ignite the peninsula, the fluidity of the situation makes prediction foolhardy.

Finally, the lack of association between anti-nuclear activism and nationalism in Russian regions of the Russian Federation provides insight into the distinctive nature of the Russian national identity. During the perestroika period, anti-nuclear activists in Russia were unable to come to grips with a distinctive national identity that could be used in their battle against Moscow. While other republics were able to use the anti-nuclear cause to demonstrate Moscow's imperial disregard for its colonies, Russians found it much more difficult to juxtapose themselves against Moscow. After all, it had been the Russian revolution that had brought communism to the region. Russia was the core of the USSR, and to many, the two entities were synonymous. Interestingly enough, while the anti-nuclear movements of Russia triggered widespread debate about the proper relationship between state and society and consideration of the degree of territorial autonomy that should be incorporated into the new order, they did not give rise to widespread discussion on what it meant to be Russian. While Yeltsin eventually incorporated anti-nuclear demands into his Russian sovereignty platform, it is important to note that the anti-nuclear activists at the local level never made this connection.

These cases illuminate important differences in the nature and strength of national identity in Armenia, Lithuania, Ukraine, Russia, and the national enclaves of Tatarstan and Crimea. While in some cases little contestation of national identity was required before a consensus was achieved among intellectual elites, in others the issues were heatedly debated and may still be far from resolution. These differences may be traced to numerous factors, including: the length of time incorporated in the USSR, method of incorporation, republic size, degree of ethnic homogeneity, preexisting histories of independent statehood, and cultural distinctiveness. In addition, however, it must be noted that the activities of movement entrepreneurs and key political figures in shaping the discourse on national identity in each region, play a significant and completely unpredictable role in determining whether a consensus might be reached on the nature of the national community.

Nuclear Power and the Public: *Looking to the Future*

During their heyday, the anti-nuclear movements of Russia, Ukraine, Lithuania, and Armenia proved remarkably successful in curtailing nuclear power construction in the region. From 1988 to 1991, the Soviet government, finding itself under the unexpected and unaccustomed onslaught of an angry public, agreed to the suspension, cancellation, or closure of over fifty nuclear reactors across the entire USSR. Confronted by mass opposition as well as the insubordination of city, oblast, and republic soviets who supported the anti-nuclear activists, the USSR government simply turned tail and ran. Never before having to take public opinion into account in their economic planning, Soviet decision makers found themselves unprepared to deal with mass opposition and were frightened by its apparent power. Interestingly enough, the government's long history of excluding the public from the decision-making process seems to have accentuated rather than limited the impact of the explosion of popular opposition to nuclear power that occurred in the late perestroika period.

After 1991, however, the influence of the anti-nuclear power movement in the Soviet successor states waned significantly. While both Russia and Ukraine, the Soviet Union's nuclear power flagships, had adopted five-year moratoria on the construction of new nuclear facilities in 1990, by late 1992 the new decision makers in these now independent states were working to distance themselves from these earlier decrees. The Russian and Ukrainian parliaments rapidly overturned the moratoria during 1992–93, and construction on a number of important unfinished nuclear projects was quick to resume. In 1993, the Russian government announced plans for the expansion of several nuclear power stations; in 1994, the Ukrainians followed suit with a decree to expand the Khmelnitsky, Zaporozhoye, and Rovno facilities.[1] Even bolder was the Armenian government's decision to reopen the Medzamor station, closed in 1988 following the immense earthquake in the region. The resolve of the Russian, Ukrainian, and Armenian governments to move ahead with their nuclear power expansion programs contrasts sharply with the dithering of the Soviet government during its final years. What might explain this change in attitude?

A number of interconnected factors may have contributed to the reversals in nuclear power policies after 1991. One of the most important of these was the elimination of a largely illegitimate form of government and its replacement with more popularly supported political institutions. Not only did these new states identify with a more deeply rooted sense of national community than that of the old USSR but they were also built upon democratic principles. Because democracies provide opportunities for the routine incorporation of public opinion in decision making, the governments are much more likely to be accepted as legitimate, and governmental decisions are less vulnerable to popular challenge. The new confidence of the governments of the Soviet successor states was immediately apparent as protesters in 1992 and after found themselves largely ignored by their governments. As these new democracies consolidated their authority, demonstrations and "meetings" became ever less effective tools for shaping political decisions.

In addition to the decreasing effectiveness of the mass-protest tactic, reversals on nuclear power policy can also be traced to changing public opinion on this issue. This is particularly true of political figures but also extended to the public at large. For politicians who had once used the nuclear power issue as a bludgeon against their old Soviet overlords, nuclear power took on new meaning once the Soviet yoke had been thrown off. While opposition to nuclear power had a highly symbolic role in the late perestroika period, after 1991 decision makers in the Soviet successor states were forced to consider the real consequences of halting their nuclear power programs. With shrinking GNP and energy shortages abounding (particularly in Armenia and Ukraine), the wisdom of closing viable stations or discontinuing construction on nearly completed projects became highly questionable. From a pragmatic point of view, the completion of any reactors near operational became both economically advantageous and desirable.

The larger public also seemed more aware of the practical consequences of opposing nuclear facilities. With evidence of economic crisis confronting them at every turn and shortages in electricity often directly affecting people's home and work lives, the symbolism of the anti-nuclear crusade was largely stripped away. The stark reality of cold, dark winters and halted industrial facilities loomed ominously. No longer did opposition to nuclear power mean opposition to Moscow's imperial domination; instead it meant undermining attempts of the newly independent state to survive economically, and it implied direct personal hardships. As a result, during 1992 and 1993, widespread popular opposition to nuclear power virtually disappeared across all of the former USSR. The Russian and Ukrainian decisions to cancel their nuclear moratoria brought statements of indignation from environmental organizations in Moscow and Kiev but failed to ignite the mass protests observed several years before.

In announcing modest expansion programs in Russia and Ukraine in 1993–94, government decision makers were apparently limited not by public opinion but rather by financial considerations. Due to the poor state of their economies, neither government could afford to launch an ambitious expansion program. While construction of nearly completed reactors may be financially feasible, initiation of new projects is not. While plans for new stations and reactors are under discussion in Russia and Ukraine (and even Armenia), it is widely acknowledged that current financial conditions do not permit adoption of such programs. A review of the Russian and Ukrainian press for 1994–1995 shows enormous problems in even keeping currently operating nuclear power stations running. Due to lack of payments by both governments and industrial users of electricity, almost all of the stations are reporting huge shortfalls in earnings. In a recent news conference reviewing the state of nuclear power in

Ukraine, chairman of the State Nuclear Committee, Mykhaylo Umanets warned that,

> the branch is on the verge of bankruptcy and technical incapacity. The spring–summer season [1995] of repair work is endangered. There are neither spare parts nor nuclear fuel to conduct repairs. Programs for improving the safety of power units are virtually no longer financed.[2]

These cash crises have also resulted in inability to purchase new nuclear fuel to keep the stations running and in failure to pay nuclear station workers. Whereas the nuclear power protests of the late perestroika period were by opponents of nuclear power among the public, the protests of the 1990s are by the nuclear power workers![3] At station after station, workers have demonstrated and held strikes in protest of the stations' failure to pay wages—often for months at a time.

Still, while mass opposition to nuclear power has dwindled since 1991, there is no reason to give up hope that an effective environmental movement may emerge to pressure these new governments to consider nuclear safety. While the symbolic function of these movements has been largely eliminated, the nuclear power problem still remains. Thus anti-nuclear clubs and organizations that emerge now are likely to more accurately represent the environmental and safety concerns of their members than the earlier politicized movements of the perestroika period. Anti-nuclear and environmental activism is entering a new era, with new leaders, forms of organization, and tactics and within a new political context. With a more professional profile and sophisticated legal and lobbying tactics, environmental organizations have the potential to play an increasingly important role in shaping nuclear power and environmental policy in the Soviet successor states. The question still remains, however, as to the extent to which these new organizations will succeed in shaping popular opinion and mobilizing citizen concern and activism on a broad scale.

NOTES

Introduction: Anti-nuclear Activism in Comparative Perspective

1. Joseph Morone and Edward Woodhouse, *The Demise of Nuclear Energy? Lessons for the Democratic Control of Technology* (New Haven, Conn.: Yale University Press, 1989).
2. For a comprehensive review of anti-nuclear movements around the globe, see Wolfgang Rudig, *Anti-Nuclear Movements: A World Survey of Opposition to Nuclear Power* (Essex, U.K.: Longman Group, 1990).
3. For a highly convincing presentation of this argument in the case of France, see Dorothy Nelkin and Michael Pollak, *The Atom Besieged: Extraparliamentary Dissent in France and Germany* (Cambridge: MIT Press, 1981). The Japanese case is discussed in David Apter and Nagayo Sawa, *Against the State: Politics and Social Protest in Japan* (Cambridge: Harvard University Press, 1984); and Margaret McKean, *Environmental Protest and Citizens' Movements in Japan* (Berkeley: University of California Press, 1981).
4. Kenneth Jowitt, "Soviet Neotraditionalism: The Political Corruption of a Leninist Regime," *Soviet Studies* (July 1983): 275–97; and Andrew Walder, *Communist Neotraditionalism: Work and Authority in Chinese Industry* (Berkeley: University of California Press, 1986).
5. Excellent histories and analyses of the events surrounding the Chernobyl disaster abound. See for example Grigorii Medvedev, *The Truth about Chernobyl* (New York: Basic Books, 1991); David Marples, *Chernobyl and Nuclear Power in the USSR* (New York: St. Martin's Press, 1986) and *The Social Impact of the Chernobyl Disaster* (New York: St. Martin's Press, 1988); D. J. Peterson, *Troubled Lands: The Legacy of Soviet Environmental Destruction* (Boulder, Colo.: Westview Press, 1993); and William Potter, "The Impact of Chernobyl on Nuclear Power Safety in the Soviet Union," *Studies in Comparative Communism* 24, no. 2 (June 1991): 191–210.
6. Jane Dawson, "Intellectuals and Anti-Nuclear Protest in the USSR," in *Beyond*

Sovietology: Essays in Politics and History, ed. Susan Solomon (Armonk, N.Y.: M.E. Sharpe, 1993), 94–124.

7. While the cancellation was never officially confirmed by the USSR Council of Ministers, an official commission of the USSR Academy of Sciences recommended cancellation in August 1988 and construction was halted at that time. See *Izvestiya* (Aug. 31, 1988).

8. Closure reported in the Armenian journal *Kommunist* (Oct. 8, 1988).

9. For a list of projects canceled or suspended during the 1986–93 period, see Dawson, "Intellectuals and Anti-Nuclear Protest."

10. Decree signed by Boris Yeltsin on June 28, 1990, *Associated Press* (June 29, 1990).

11. *Pravda Ukraina* (Aug. 8, 1990).

12. The Russian government approved reinitiation of their nuclear power program in October 1992; see *Nuclear Engineering International* (Feb. 1993): 3. The following year, Ukraine's parliament overturned its nuclear moratorium; see *Nuclear Engineering International* (Dec. 1993): 2.

13. Reported by *Radio Mayak* (Jan. 26, 1995).

14. Jane Dawson, "Anti-nuclear Activism in the USSR and Its Successor States: A Surrogate for Nationalism?" *Environmental Politics* 4, no. 3 (1995): 441–66.

15. While the impact of this movement on nuclear power decision making during the perestroika period certainly indicates a dramatic change in the way government decisions were made under Gorbachev, this is not a study of the decision-making process. Rather, my goal is to add to our comparative understanding of how and why popular opposition to nuclear power has been mobilized around the globe by providing a detailed study of anti-nuclear activism in the late- and post-communist societies of Russia, Ukraine, and Lithuania.

16. This structural interpretation relies heavily on insights drawn from Resource Mobilization Theory. For excellent reviews of this theoretical literature, see J. Craig Jenkins, "Resource Mobilization Theory and the Study of Social Movements," *Annual Review of Sociology* 9 (1983): 527–53; and Aldon Morris and Cedric Herring, "Theory and Research in Social Movements: Critical Review," in *Political Behavior Annual,* ed. Samuel Long (Boulder, Colo.: Westview Press, 1984).

17. This point is based on McCarthy and Zald's distinction between professional and classical social movement organizations. See John McCarthy and Mayer Zald, "Resource Mobilization and Social Movements," *American Journal of Sociology* 82 (1977): 1212–41.

18. While other environmental causes have been linked to nationalism outside the USSR, in the developed Western world the linkage between the anti-nuclear platform and nationalism has not been prominent.

19. Because conditions did not permit field research in Armenia from 1990 to 1991, the Armenian case is not treated in detail in this study. A brief overview of anti-nuclear activism in Armenia is included, however, in chapter three.

1. Patterns of Social Mobilization in Late- and Postcommunist Societies

1. For excellent reviews of the resource mobilization literature, see J. Craig Jenkins, "Resource Mobilization Theory and the Study of Social Movements," *Annual Review of Sociology* 9 (1983): 527–53; and Aldon Morris and Cedric Herring, "Theory and Research in Social Movements: Critical Review," in *Political Behavior Annual,* ed. Samuel Long (Boulder, Colo.: Westview Press, 1984).
2. Work on social movements by Alain Touraine, Albert Melucci, and Alessandro Pizzorno epitomize this school. See for example Touraine, *The Voice and the Eye* (New York: Cambridge University Press, 1981); Melucci, "The Symbolic Challenge of Contemporary Movements," *Social Research* 52, no. 4 (winter 1985): 789–816; and Pizzorno, "Political Exchange and Collective Identity in Industrial Conflict," in *The Resurgence of Class Conflict in Western Europe Since 1968,* ed. Crouch and Pizzorno (London: Macmillan, 1978), 277–98.
3. A hybrid approach has been suggested by Jean Cohen, "Strategy or Identity: New Theoretical Paradigms and Contemporary Social Movements," *Social Research* 52, no. 4 (winter 1985): 663–716; Doug McAdam, John McCarthy, and Mayer Zald, "Social Movements," in *Handbook of Sociology,* ed. Neil Smelser (Newbury Park, Calif.: Sage Publications, 1988); Albert Melucci, "Getting Involved: Identity and Mobilization in Social Movements," in *International Social Movement Research,* ed. B. Klandermans, H. Kriesi, and Sidney Tarrow (Greenwich, Conn.: JAI Press, 1988); Jurgen Gerhards and Dieter Rucht, "Mesomobilization: Organizing and Framing in Two Protest Campaigns in West Germany," *American Journal of Sociology* 98, no. 3 (1992): 555–596; Sidney Tarrow, "Mentalities, Political Cultures, and Collective Action Frames: Constructing Meanings through Action," in *Frontiers in Social Movements Theory,* ed. Aldon Morris and Carol Mueller (New Haven, Conn.: Yale University Press, 1992).
4. Leading examples of this approach include Neil Smelser's collective behavior, Ted Gurr's relative deprivation, and William Kornhauser's mass society approaches. Smelser, *Theory of Collective Behavior* (New York: Free Press of Glencoe, 1963); Gurr, *Why Men Rebel* (Princeton, N.J.: Princeton University Press, 1970); Kornhauser, *The Politics of Mass Society* (New York: Free Press of Glencoe, 1959).
5. John McCarthy and Mayer Zald, *The Trend of Social Movements* (Morristown, N.J.: General Learning, 1973), 13.
6. Mancur Olson, *The Logic of Collective Action* (New York: Schocker, 1968).
7. This is the explicit task of contributors to *Frontiers in Social Movement Theory.*
8. While resource mobilization theorists initially focused on the role of selective incentives in mobilizing individuals to participate (see for example Frohlich, Oppenheimer, and Young [1971] and Oberschall [1973]), later empirical studies tended to challenge this focus on by-product benefits and support the hypothesis that ideal interests play a dominant role in the mobilization of many social move-

ments (see for example Tillock and Morrison [1979] and Berry [1977]). N. Frohlich, J. Oppenheimer, and O. Young, *Political Leadership and Collective Goods* (Princeton, N.J.: Princeton University Press, 1971); Anthony Oberschall, *Social Conflict and Social Movements* (Englewood Cliffs, N.J.: Prentice-Hall, 1973); H. Tillock and D. Morrison, "Group Size and Contributions to Collective Action," *Research in Social Movements* 2 (1979); Jeffrey Berry, *Lobbying for the People* (Princeton, N.J.: Princeton University Press, 1977).

9. The absence of a public realm is a key feature in the neotraditional (see Jowitt [1983], Walder [1986]), totalitarian (see Friedrich and Brzezinski [1965]), and post-totalitarian models (see Kassof [1964], Meyer [1977]) of communist state-society relations. Even more pluralist approaches such as Hough's model of institutional pluralism acknowledge the lack of opportunities for independent public activities as a continuous feature of communist systems. Kenneth Jowitt, "Soviet Neotraditionalism: The Political Corruption of a Leninist Regime," *Soviet Studies* (July 1983: 275–297); Andrew Walder, *Communist Neotraditionalism: Work and Authority in Chinese Industry* (Berkeley: University of California Press, 1986); Carl Friedrich and Zbigniew Brzezinski, *Totalitarian Dictatorship and Autocracy* (Cambridge, Mass.: Harvard University Press, 1965); Allen Kassof, "The Administered Society: Totalitarianism without Terror," *World Politics* (July 1964: 558–575); M. Meyer, "USSR Incorporated," *Slavic Review* 20 (1961); Jerry Hough, *The Soviet Union and Social Science Theory* (Cambridge, Mass.: Harvard University Press, 1977): 369–376.

10. Typology of resources suggested by John Freeman, "Resource Mobilization and Strategy," in *The Dynamics of Social Movements,* ed. Mayer Zald and John McCarthy (Cambridge, Mass.: Winthrop, 1979).

11. Walder, *Communist Neotraditionalism: Work and Authority in Chinese Industry.*

12. Philip Selznick, *The Organizational Weapon: A Study of Bolshevik Strategy and Tactics* (New York: McGraw-Hill, 1952).

13. See for example Gorbachev's speeches to the October 1985 and January 1987 Communist Party of the Soviet Union (CPSU) Central Committee Plenums.

14. Text of "Law on Public Associations" in *Pravda* (Oct. 16, 1990): 3.

15. For an excellent discussion of the Soviet Union's egalitarian welfare structure as well as the efficiency shortcomings which plagued the Soviet economy, see Ed Hewett, *Reforming the Soviet Economy: Equality versus Efficiency* (Washington, D.C.: The Brookings Institution, 1988).

16. While the annual rate of growth in GNP for the period 1928–55 has been estimated at 4.4–6.4% (Bergson [1961]), growth rates declined steadily throughout the 1970s and early 1980s. Hewett (1988) provides the following statistics for average annual GNP growth rates: 4.7% (1961–64), 5.0% (1966–70), 3.0% (1971–75), 2.3% (1976–80), 2.0% (1981–85). Economic performance continued to decline during the Gorbachev period; growth in GNP fell from 2.5% (1985–86) to 1.7% (1987–88) to 1.4% (1989) and finally moved into the nega-

tives in 1990 (Noren [1991]). Following the breakup of the USSR, 1992–94, most of the Soviet successor states continued to exhibit negative growth patterns. A. Bergson, *The Real National Income of Soviet Russia since 1928* (Cambridge: Harvard University Press, 1961), 261, cited in Hewett, *Reforming the Soviet Economy,* 38; Hewett, *Reforming the Soviet Economy,* 52; J. Noren, "The Economic Crisis: Another Perspective," in *Milestones in Glasnost and Perestroyka: The Economy,* ed. Ed Hewett and Victor Winston (Washington, D.C.: The Brookings Institution, 1991), 360–407.

17. See for example Frances Piven and Richard Cloward, *Poor People's Movements: Why They Succeed and How They Fail* (New York: Pantheon, 1977).

18. Based on McCarthy and Zald's distinction between professional and classical SMOs. See McCarthy and Zald, "Resource Mobilization and Social Movements," *American Journal of Sociology* 82 (1977): 1212–41.

19. Piven and Cloward, *Poor People's Movements: Why They Succeed and How They Fail.*

20. Charles Tilly, *From Mobilization to Revolution* (Reading, Mass.: Addison-Wesley, 1978).

21. This is consistent with studies on impoverished peasant societies that view maintenance of minimal subsistence levels as their key objective and that mobilize defensively whenever these minimal levels appear threatened from without. See J. Scott, *The Moral Economy of the Peasant* (New Haven, Conn.: Yale University Press, 1976).

22. Note that this does not preclude offensive mobilization by poor constituencies under certain circumstances. First, while poor groups are unlikely to mobilize offensively on their own, they may choose to do so when the mobilization is orchestrated by experienced movement organizers and financed by other social groups. That is, the potential exists for "movement entrepreneurs" to mobilize groups that might otherwise remain dormant. (McCarthy and Zald, "Resource Mobilization and Social Movements.") Secondly, while poor constituencies are initially prone toward defensive mobilizations, the possibility exists that if defensive objectives are met, the goals of the movement may expand in offensive directions.

23. Note that the most significant shift in decision-making power to constituencies outside the Communist elite only occurred in the spring of 1990, when multicandidate elections were held for local offices across the USSR. The shift in power, however, continued to be limited by restrictions on nominations by independent associations and the party's control over the registration of candidates. Furthermore, in many cases these newly elected officials found themselves confronted by entrenched ministerial and party bureaucracies which diminished the power of the quasi-popularly elected bodies. Thus, while participatory channels were expanded from 1985 to 1991, it seems fair to say that groups outside the Communist nomenklatura remained in a highly disadvantaged power position relative to the party and state.

24. This argument is particularly well presented in Philip Roeder's article, "Soviet Federalism and Ethnic Mobilization," *World Politics* 43, no. 2 (Jan. 1991): 196–232.
25. The Baltic countries seem to be an exception to this trend.
26. This is also true of non-Russian enclaves within the borders of the Russian Federation such as Tatarstan.
27. In addition, as the national movements began to emerge in non-Russian regions of the USSR, they tended to quickly eclipse anti-nuclear activities. As the national movements gained confidence, they quickly began to leave their defensive agendas behind and move in offensive directions. By 1990, many were calling not only for the protection of their nation but also for sovereignty and eventual independence. This shift from defensive to offensive mobilizational platforms will also need to be considered more thoroughly in the case studies that follow.

2. Lithuania: The National Element

1. Two types of nuclear reactors predominated in the Soviet nuclear power sector: the VVER (water-cooled) and the RBMK (graphite-moderated). RBMK reactors were (and continue to be) used at the Chernobyl nuclear power station. These reactors have no containment, and their cooling system is considered by many to be inferior to reactors commonly found in the United States and Western Europe. Power output from these immense RBMK reactors ranges from 1,000 to 1,500 MW.
2. Judith Thornton, "Soviet Electric Power after Chernobyl: Economic Consequences and Options," *Soviet Economy* 2, no. 2 (1986): 131–79.
3. Because little information has been published on the scientific debate over the siting and expansion of Ignalina AES prior to 1986, this section relies heavily on in-depth interviews with scientists who were directly involved in this process. Algirdas Žukauskas, vice president of the Lithuanian Academy of Sciences, was particularly helpful in providing details from this period. His historical account was corroborated through interviews with geologist Povilas Suveizdis of the Lithuanian Academy of Sciences' Institute of Geology and physicists (and leading members of Lithuanian's first anti-nuclear association) Vigmas Vaišvila, Jonas Tamulis, and Aidas Vaišnoras of the Lithuanian Academy of Sciences' Institute of Physics (spring 1991).
4. This factor was particularly stressed in interviews with Povilas Suveizdis, senior geologist of the Lithuanian Academy of Sciences' Institute of Geology (spring 1990).
5. Interview with Algirdas Žukauskas, vice president of the Lithuanian Academy of Sciences (June 1991).
6. J. Jablonskis, R. Janukėnienė, and M. Lasinskas, "Elektrinese pašildyto vandens poveikis aplinkai," *Mosklas ir technika* 9 (1980): 46, cited in Augustine

Idzelis, "The Socioeconomic and Environmental Impact of the Ignalina Nuclear Power Station," *Journal of Baltic Studies* XIV, no. 3 (fall 1983): 247–54.

7. While documentation on these closed and highly secretive sessions is difficult to come by, I received access to one report on an interdepartmental conference on Ignalina held at the Institute of Power Engineering (NIKIET: Nauchno-issledovatel'skii i konstruktorskii institut energotekhniki) (July 19, 1983). The conference concluded that Lake Druksai could sustain a maximum of 3,000–4,500 MW of power production.

8. Idzelis, "The Socioeconomic and Environmental Impact."

9. S. Voronitsyn, "The Chernobyl Disaster and the Myth of the Safety of Nuclear Power Stations," *RL Research Bulletin* 202 (1986).

10. Exceptions include a 1979 article by N. Dollezhal and Yurii Koryakin, and a number of local articles highlighting construction delays and flaws at numerous reactors across the USSR. It should be noted, however, that this latter category of criticism was limited to labor discipline and never attacked nuclear power in general. See for example the *Radio Vilnius* report of Nov. 1, 1984, which complained of construction delays and other problems at the Ignalina AES. Dollezhal and Koryakin, "Yadernaya energetika: Dostizheniya i problemi," *Kommunist* 14 (Sept. 1979): 19–28.

11. Note the April 1987 article published in Sweden claiming cancellation of #3 and #4, and the USSR's lack of response to this article. Reported in *RL Research Bulletin* (May 4, 1988). The official announcement came only in April 1988; see *Tiesa* (Apr. 1, 1988).

12. This finding is based on dozens of interviews with former *druzhina* members in Lithuania, Russia, and Ukraine.

13. This finding is supported by an extensive search through the Samizdat Archives of Radio Liberty, Munich, Germany (fall 1992). This archive of unofficial and underground East Bloc publications represents one of the most extensive collections of these materials in the world. Two weeks of searching, however, yielded no Lithuanian dissident writings on the nuclear power issue. (Note that these archives have recently been transferred to the Open Media Research Institute in Prague).

14. See my discussion of glasnost and the nuclear power issue in "Intellectuals and Anti-Nuclear Protest in the USSR," in *Beyond Sovietology: Essays in Politics and History,* ed. Susan Solomon (Armonk, N.Y.: M.E. Sharpe, 1993), 94–124.

15. This fact was noted by S. Girnius in "The Ignalina Atomic Power Plant's Second Reactor in Operation," *RL Research Bulletin* (May 4, 1988).

16. This was noted by S. Girnius in *RL Research Bulletin* (July 1, 1986).

17. This history of the Žemyna association is based primarily on my in-depth interviews with early members of the club's leadership, including Zigmas Vaišvila, Aidas Vaišnoras, Jonas Tamulis, Alvydas Medalinskas, A. Dautaras, and Arturas Abromavičius (spring 1991).

18. For example, mathematician Statulavichas.

19. While a handful of writers and artists were members of Žemyna, the club was primarily composed of physical scientists, particularly physicists and mathematicians. Many of these young scientists had become acquainted with each other as students studying in Moscow several years previously.
20. Note the progressive orientation of the Lithuanian Komsomol in sheltering the group and publishing its articles.
21. *Tiesa* (Mar. 24, 1988) and (Apr. 1, 1988).
22. This history of the formation of the Lithuanian Movement for Perestroika, or Sajudis, is drawn primarily from interviews with Sajudis spokesman Andreus Kuibilus and early Sajudis members Zigmas Vaišvila, Jonas Tamulis, and Alvydas Medalinskas (all members of Žemyna as well) and Vytautas Bogušis (later member of the more radical Lithuanian Freedom League). It is supported by the chronology presented in *Sajudis: A Brief History* (published by Sajudis) as well as reports by Senn and Krickus. See Albert Senn, *Lithuania Awakening* (Berkeley: University of California Press, 1990); Richard Krickus, "Lithuania: Nationalism in the Modern Era," in *Nations and Politics in the Soviet Successor States,* ed. Ian Bremmer and Ray Taras (Cambridge: Cambridge University Press, 1993).
23. For more details on the emergence of informal groups during the spring of 1988, see Senn, *Lithuania Awakening.*
24. In interviewing anti-nuclear activists in Lithuania, I constantly heard the Ignalina and Sniečkus problems referred to as "demographic pollution," thus indicating that hostility to the nuclear power station went far beyond mere scientific concerns.
25. See *Komjaunimo tiesa* (July 23, 1988) report on a meeting of 300 people held at Ignalina to protest the construction of reactor #3. The meeting was organized by 115 cyclists who were traveling around Lithuania to publicize environmental issues. The bicycle trip was sponsored by Sajudis and Žemyna.
26. See also Brazauskas's interview in *Tiesa* (June 23, 1988). In this interview, Brazauskas notes that final construction plans for reactor #3 were never confirmed and approval was not given by either the Lithuanian water resources agencies or the USSR Atomic Control Commission. He thus questions whether construction of #3 should be going ahead.
27. Other public rallies against Ignalina include an unofficial gathering of approximately two hundred people at the Lithuanian Supreme Soviet on July 15 specifically to protest construction at Ignalina and a major rally of five hundred thousand people on August 2 that focused on environmental issues, among other topics.
28. *Tiesa* (Aug. 18, 1988).
29. This figure was given by Kromchenko at roundtable (Aug. 25, 1988).
30. This decision was also published in *Izvestiya* (Aug. 31, 1988). This article focuses on design flaws associated with the failure to take into account the seismicity in the area. It also hints that these problems may affect the first two reactors as well.

31. See announcement of recent fires in *Izvestiya* (Sept. 5, 1988). Earlier fires occurred in July but were not immediately reported to the public.
32. See appeal made by the Lithuanian Freedom League on September 5 and interview with Bogušis (Reuters, Sept. 14), leading spokesperson for the League.
33. Videotapes of the weekend events include interviews with residents of Sniečkus and Ignalina that reflect anger, resentment, and a complete lack of sympathy with the protestors' cause. (Tape available through Aidas Vaišnoras, the Ecological Center Alternatyva, Vilnius, Lithuania.)
34. The work and results of this commission were reported in detail in a Radio Vilnius report to North America (3 January 1989).
35. Jonas Tamulis, member of the Ecology Commission of the Lithuanian Supreme Soviet, interview by author (June 1991).
36. Reported in *Nuclear Engineering International* (Mar. 1994): 2.
37. Because conditions did not permit me to travel to Armenia in 1990 or 1991, this section is based largely on a series of interviews with environmental activists in Armenia conducted by Russian sociologist Olga Lobach of the USSR Academy of Sciences Institute for Nuclear Safety (fall 1991).
38. Elizabeth Fuller, "Armenian Authorities Appear to Yield to Ecological Lobby," *RL Research Bulletin* 130 (1987).
39. The percentage of ethnic Armenians living in Armenia was reported as 93.7% in 1989; see Mouradian, *De Staline a Gorbatchev* (1990), cited in *Nations and Politics in the Soviet Successor States,* p. 269.
40. Closure decision reported in *Kommunist* (Armenian) (Oct. 8, 1988). Note that while some have attributed this closure to concerns about seismicity, Dr. Sergei Pankratov and members of his assessment team who provided the recommendation for closure maintain that concerns about public acceptability were the primary factors in the closure decision. Contrary to popular belief, the closure decision actually predates the December 1988 earthquake. Dr. Pankratov and members of his team, interview by author, USSR Academy of Sciences Institute of Nuclear Safety (Mar. 1990).
41. President Ter-Petrosyan reiterated his support for reopening the Medzamor station in a report on the overall socioeconomic situation in Armenia in early 1993. Noted in *Nuclear Engineering International* (Feb. 1993): 3.
42. *Respublika Armeniya* (Feb. 1, 1995): 1.
43. This is according to Armenian Fuel and Power Minister, Miron Shishmanyan. Reported by *Radio Mayak* (Jan. 26, 1995).

3. Ukraine: Civic or Ethnic Mobilization?

1. These include four RBMK-1000 reactors at Chernobyl AES, two VVER-1000 reactors at Zaporozhoye AES, two VVER-1000 reactors at South Ukraine AES, and two VVER-440 reactors at Rovno AES.

2. Judith Thornton, "Soviet Electric Power after Chernobyl: Economic Consequences and Options," *Soviet Economy* 2, no. 2 (1986): 131–79.
3. New reactors include two RBMK-1000's at Chernobyl AES, four VVER-1000's at Chigirin AES, two VVER-1000's at Rovno AES, two VVER-1000's at South Ukraine AES, one VVER-1000 at Crimean AES, four VVER-1000's at Khmelnitsky AES, two VVER-1000's at Kharkov nuclear heating stations (ATETS), two VVER-1000's at Odessa ATETS, and two VVER-1000's and two VVER-1500's at Zaporozhoye AES.
4. Yurii Shcherbak, interview by author (Feb. 1990). Dr. Shcherbak, founding member of the environmental organization Zelenii svit, has served as Ukrainian minister of the environment and is currently Ukraine's ambassador to the United States.
5. Yurii Shcherbak, interview by author (spring 1990).
6. Jane Dawson, "The Soviet Scientific-Technical Intelligentsia: An Emerging Public?" in *Analyzing the Gorbachev Era,* ed. Gail Lapidus (Berkeley, C.A.: Berkeley-Stanford Publication, 1989), 87–116.
7. See for example Oles Honchar's article in *Literaturna Ukraina* (Oct. 7, 1987) in which he criticizes the nuclear power program in Ukraine and asks which of Ukraine's other reactors are actually hidden Chernobyls waiting to happen.
8. See for example the collective letter of seven writers asking for the cancellation of Chigirin AES, published in *Literaturna Ukraina* (Aug. 1987).
9. Dr. Shcherbak was one of the first writers to expose the incompetence of the authorities in evacuating the population during the early days following the Chernobyl disaster. See his *Chernobyl: A Documentary Story* (London: Macmillan Press, 1989).
10. Note that, unlike in Lithuania, this scientific opposition was not supported or sheltered by the Ukrainian Academy of Sciences. Interviews with leading members of the Ukrainian Academy reveal that the Academy played no role in this conference.
11. K. Grigorev and S. Kiselev, *Literaturnaya gazeta* (May 27, 1987).
12. While no reports of these meetings appeared in the Ukrainian press at the time, details of the August 25 meeting (as well as other meetings) and its results appeared in a joint letter of thirteen leading scientists which was published in *Literaturna Ukraina* (Jan. 1988). The thirteen scientists strongly opposed the expansion of the three above-mentioned Ukrainian AESs and complained that the USSR Ministry of Atomic Energy had completely disregarded the recommendations of the August 25 session.
13. *TASS* announced on December 31, 1987, that the Khmelnitsky AES had begun operation. Noted in *Foreign Broadcast Information Service (Soviet Union)* (Jan. 4, 1988): 46.
14. One of the most active members of this small group was Panov. In 1987, Panov took the daring step of publishing an article in *Literaturna Ukraina* entitled, "Nuzhna li nam atomnaya energetika dal'she?"

15. At this conference, writers demonstrated a new strategy of utilizing scientific data to justify their stance on nuclear energy. Poet Ivan Drach noted that scientists had already calculated the maximum nuclear capacity sustainable on the remaining appropriate sites in Ukraine and had concluded that this maximum would be reached when the reactors currently under construction were finished, thus making any further expansion of nuclear power in Ukraine unadvisable.
16. Note that many of these local activists outside Kiev found themselves harassed and occasionally persecuted for their participation in independent antinuclear activities. In interviews with local activists involved in the anti-Khmelnitsky movement, I learned that all of them had been called in for questioning by the KGB in regard to their activities and warned to cease and desist. Because antinuclear sentiment was so strong among the population, however, local officials generally resisted taking severe actions against movement leaders. The power station administrators, however, were less sympathetic to the movement, and workers who assisted the movement during this early period often found themselves unemployed.
17. Note that plans to construct nuclear power and heating stations (ATETs) near Odessa, Kharkov, and Kiev were canceled before significant popular opposition emerged. These cancellation decisions were based on the unusual proximity of these stations to large population centers. Also, protests against the expansion of Zaporozhoye AES did not begin until much later (1990); eastern Ukraine is predominantly Russian, and activation of this region occurred much later than other regions of Ukraine and was much weaker.
18. The draft statutes necessary for registration were decided at a gathering of leading Zelenii svit activists in May 1989. At this meeting, the organizational structure of Zelenii svit was established, and according to numerous reports, the organization finally began to function.
19. This firing is substantiated by numerous letters that Zelenii svit sent to the government protesting the unfair treatment of this activist.
20. Note that, as an umbrella organization, Zelenii svit only permitted groups to register. No individual memberships in Zelenii svit were allowed.
21. A. Glazovoi and A. Abdullen of *Rabochaya gazeta,* interview by author, Kiev (spring 1990).
22. Published in *Literaturna Ukraina* (Feb. 16, 1989).
23. See for example *Pravda* (June 21, 1989).
24. This was noted in my interview with Myroslav Popovich (spring 1990), as well as in Roman Solchanyk's interview with Pavlo Movchan in *Ukraine: From Chernobyl to Sovereignty* (New York: St. Martin's Press, 1992).
25. Reported in *Radyanska Ukraina* (Apr. 2, 1989).
26. Published in *Pravda Ukrainy* (Aug. 8, 1990).
27. Reported in *Nuclear Engineering International* (Dec. 1993).
28. *ITAR-TASS* Report (May 12, 1994).

4. The Battle against Khmelnitsky AES: A Close-up View of Mobilization in Ukraine

1. The Khmelnitsky oblast was one of only seven oblasts with more than half of its population living in the countryside. The majority population of the other 18 oblasts lived in urban centers. "Territory and Population by Oblast on Jan. 1, 1990," in *Narodnoye khozyaistvo Ukrainskoi SSR: 1989* (Kiev: Tekhnika, 1990): 28.
2. These attempts to circulate petitions were discussed at a roundtable of informal group representatives held in Shepetovka during the spring of 1991.
3. Note that not only were there no anti-nuclear informal groups in the oblast at that time, there is no evidence of any kind of informal organizations in the region in 1987.
4. See, for example, the information on contamination levels in Zhitomir oblast which became public for the first time in 1989. Exposés were published by V. Kosarchuk and I. Petrenko in *Molod' Ukrainy* (Apr. 19, 20, 1989), cited in D. Marples, *Report on the USSR* (May 19, 1989).
5. For a detailed review of the government's ineptitude in evacuating workers and residents after the accident, see Yurii Shcherbak, *Yunost'* 6 (1987); 9 (1987): 5–16; 10 (1987): 11–29.
6. The seven points of the Nabat platform can be found in the program of the March 15 meeting that the group prepared afterwards. Points include (1) an immediate halt in construction at Khmelnitsky AES, (2) removal of all used fuel from the republic, (3) a roundtable between local activists, Shcherbitsky and representatives of the relevant Moscow ministries, (4) a media campaign against Khmelnitsky AES, (5) meetings and demonstrations to demand a halt in construction and complete closure of the station, (6) work to prevent local authorities from harassing informal group members, and (7) work to officially register the group. (Copy of text available in author's personal archive.)
7. This is supported by interviews with Opanasiuk and Nagulko as well as by discussions at a roundtable of informal group representatives held in Shepetovka, spring 1991.
8. I was able to meet with Matunina briefly in Kiev in 1990 and to carry out an in-depth interview with her the following year in Netishin.
9. Note that the term *meeting* was used by activists to denote a large, public rally. Thus, while the term was adopted from English, its meaning was slightly different.
10. Note that *Trudyvnik Polessa* (May 5, 1989) reports that an unsanctioned "meeting" was held in Netishin on the Chernobyl anniversary (Apr. 26, 1989). According to the newspaper report, this was the first major demonstration in the area and was viewed as a shocking event by most of the local population. While the meeting did not receive official permission, the police were careful to avoid confrontation with the protesters.
11. Note that the Ostrog-Netishin chapter of Zelenii svit ran into real problems

registering locally. Thus, the group was not officially registered until early 1990. At this time they took advantage of the privilege of the all-Ukraine Zelenii svit association to register local groups under its umbrella without permission of local authorities.

12. See *Trudyvnik Polessa* (June 30, 1989) for a detailed report on this meeting.

13. See the list of meeting resolutions, April 16, 1989 (author's private archive). An interesting response to this meeting can be found in a letter from a workers collective at the Shepetovka OTEP-46895 factory. In this letter the workers collective thanks Zelenii svit for organizing the April 16 meeting, and voices its support for halting the expansion of Khmelnitsky AES and converting the first reactor over to a conventional fuel station. The letter is signed by fifteen members of the workers collective. While I only had access to one such letter (dated April 1989), interviews with leaders of Shepetovka Zelenii svit indicate that such letters were quite common and support among workers was high.

14. Unfortunately, due to the paper shortage in the oblast, no archive of back issues was kept by the newspaper. Thus, I was not able to find copies of all issues covering the nuclear power topic. Those I found, however, were exclusively antinuclear. In addition, interviews with activists indicate that they viewed this paper as a supporter of their cause.

15. A particularly interesting example of changing communist attitudes toward the Khmelnitsky AES can be found in the outcome of the plenum of the Zdolbunovsky regional party organization, held in February 1990. At this plenum, party members voted to send a letter of complaint to the USSR Council of Ministers, the Ukrainian Supreme Soviet, and the USSR Ministry of Atomic Energy. In this letter, they complained that Khmelnitsky AES was endangering the health of their children and their territory and they resented not having been asked whether they wanted a nuclear power station in their neighborhood. They demanded that the expansion of Khmelnitsky AES be halted and a schedule be established for closing the station entirely. (Letter from First Secretary Kononchuk (Feb. 22, 1990); author's private archive.)

16. The Ukrainian parliament's declaration of a five-year moratorium on new nuclear projects was published in *Pravda Ukrainy* (Aug. 8, 1990).

17. Khmelnitsky oblast soviet resolution, "On moratorium on construction of Khmelnitsky AES" (Apr. 6, 1990); copy available in author's private archive. The text notes that the resolution was proposed by deputies Slobodyna, Mironyuk, and Okorsky, as well as the deputies of the Izyaslavsky rayon and the town of Kamyantsia-Podilsky, and members of Shepetovka Zelenii svit. Resolution includes three points: (1) to freeze construction on reactor #2 and any additional reactors, and call on the USSR and Ukrainian Councils of Ministers to carry out an ecological review on the first reactor; (2) to ask the Minister of Power, Y. K. Semenov, and the Minister of the Power Complex, V. F. Konovalov, to review the entire complex of questions associated with the freeze decision; and (3) to ask oblast soviet deputies Guselnikov and Lundishev to work wtih USSR ministries to

transfer these construction activities to other industrial sites in the oblast and to reassign station personnel.

18. Moscow Domestic Service (June 5, 1990) reports in *Foreign Broadcast Information Service (Soviet Union)* (June 20, 1990) that 15 people were still continuing a hunger strike against the construction of Khmelnitsky AES.

19. *Radio Kiev* (May 28, 1990) reported that a two-hour warning strike was held at numerous industrial enterprises in the Khmelnitsky oblast that day, and preparations were under way for a full-scale strike in the future if construction at Khmelnitsky AES was not halted.

20. This threat is noted in numerous reports, including a worried letter from the Director of the station, Victor Sapronov, to Shepetovka Supreme Soviet deputy (and member of Zelenii svit), Taras Nagulko. (Apr. 2, 1990; author's private archive.)

21. Tatyana Matunina, head of the Ostrog chapter of Zelenii svit, played a central role in organizing this action and keeping it going over the summer.

22. Protesters bitterly complained that reports in the central press (e.g., *Izvestiya* [July 15, 1990]), grossly distorted the facts by claiming that the entire station was under blockade and workers were being prevented from entering or leaving the station.

23. Copy available in author's private archive.

24. *UNIAN* (March 18, 1994) and (March 23, 1994).

25. *ITAR-TASS* (May 12, 1994).

5. Russia: The Demand for Local Self-Determination

1. These NIMBY movements were not unique to Russia. During the early perestroika period, local movements opposing industrial facilities were a common phenomenon in most of the republics of the USSR.

2. The most successful of these attempts were probably the "Ekologiya i mir" (Ecology and Peace) and "Sotsialno-ekologicheskii soyuz" (Social-Ecological Union) groups, both based in Moscow. Although these associations managed to put together a network of people throughout the entire USSR, they often had only a handful of contact people in each area, and communication between regions was almost nonexistent. The associations acted more as informational networks and Moscow lobbying groups than as umbrellas for a mass movement.

3. In 1991, I was one of the first foreigners permitted to visit Gorky. This visit was made possible by a trial agreement between the USSR Academy of Sciences (my official host in the USSR) and the KGB for 1991 to permit foreign scholars to visit Gorky.

4. See for example, articles in *Pravda* (Aug. 14, 1980) and *Atomnaya energiya* (Mar. 2, 1980).

5. Eventually, fifteen such ASTs were to be constructed in the USSR.

6. Alexandrov is quoted as calling for the construction of ATETss in "several hundred" population centers in an article by Gubarev, *Pravda* (Jan. 30, 1984).
7. *RL Research Bulletin* (Sept. 21, 1983).
8. In an article in *Pravda* (Feb. 25, 1986), the author noted that nuclear heating facilities would soon be constructed on the outskirts of Odessa, Minsk, Kharkov, and Volgograd. He also noted the innovative new design being tested in Gorky and Voronezh.
9. J. Thornton, "Soviet Electric Power after Chernobyl: Economic Consequences and Options," *Soviet Economy* 2, no. 2 (1986): 131–79.
10. A good account of the harassment of democratic activists in Gorky can be found in *Yunost'* (Dec. 1989).
11. *Tochka zreniya* (Point of View) 1 (June 1988).
12. *Ibid.*
13. It should be noted, however, that nuclear specialists and scientists from the OKBM design institute and the Gorky AST did not participate in the anti-nuclear movement.
14. Other key scientific figures opposing the Gorky AST include Academician G. G. Devyatikh, the vice director of the Institute of Chemistry of the USSR Academy of Sciences, and Gaponov-Grekov, director of the Institute of Applied Physics of the USSR Academy of Sciences. Articles in the local press include Troitsky, *Leninskaya smena* (Dec. 28, 1988); *Sotsialisticheskaya industriya* (Mar. 26, 1989); *Stroitel'naya gazeta* (Aug. 2, 1989); *Pravda* (Sept. 15, 1989); *Leninskaya smena* (Sept. 20, 1989).
15. It is important to note that this was the first rally in Gorky to draw significant numbers of people out onto the streets. Although probably less than 2,000 people attended, this was a dramatic event in the conservative town of Gorky.
16. By April 1989, numerous sources reported over 100,000 signatures collected. See for example *Sotsialistichskaya industriya* (Apr. 1989); *Den' za dnem* 7 (Nov. 1989); or *Trud* (July 28, 1989); *Radio Moscow* (July 30, 1989).
17. The *obkom* demanded that the group pay to rent facilities, but since the group was not officially registered and had no funds, they were unable to pay rent.
18. Copy of official decree available in author's personal archive.
19. The full list of members is as follows: Devyatikh, Gaponov-Grekhov, Zaburdayev, Kisarov, Kitayev, Korolev, Mazun, Mitenkov, Molozhnikov, Napalkov, Naidenko, Neiman, Nemtsov, Perevezentsev, Ryakhin, Stepanov, Tikhomirov, Troitsky, Fiolnenko, and Shkarin.
20. Information on Vinogradov's role in the movement was gleaned primarily from personal interviews with Vinogradov, Likhachev, and Troitsky in the spring of 1991.
21. Copy available in author's personal archive.
22. See report in *Trud* (July 28, 1989).
23. See for example, letter from the Borksii Stekolnii zavod. (Available in author's personal archive.)

24. A two-hour strike was staged at the all-union selection station in August 1989.
25. Copy of letter available in author's personal archive.
26. See report in *Sovetskaya Rossiya* (Nov. 25, 1989).
27. See, for example, *Nizhegorodskaya yarmarka* 10 (Mar. 10, 1991); *Stroitel'naya gazeta* (Oct. 5, 1989); *Leninskaya smena* (Oct. 3, 1989).
28. See, for example, the report on the IAEA review in *Den' za dnem* 7 (Nov. 1989).
29. The only candidates to openly profess support for the Gorky AST were those nominated by the OKBM design institute. Because their electoral district was composed primarily of OKBM employees, these candidates were able to espouse this highly unpopular viewpoint and still receive majorities in the elections.
30. Note that in 1990, the city's name was changed from Gorky to Nizhny Novgorod. In order to avoid confusion, however, I will continue to use the name Gorky throughout the chapter. Percentage provided by Communist Party official, Vinogradov, interview by author (spring 1991).
31. Note that Nemtsov's political career has skyrocketed since that time. Nemtsov is currently governor of Nizhny Novgorod oblast and even rumored to be a possible contender for the Russian presidency in 1996.
32. Unofficial copy of decision available in author's personal archive.
33. Copy of official decision available in author's personal archive.
34. In addition to meeting with key members of the relevant oblast soviet committees, Kayumov and other members of Dront also drafted a number of environmental proposals which were frequently used as the basis for developing environmental legislation by the local soviets.
35. Copy of Gorky executive committee's record of the meeting available in author's personal archive.
36. The alternatives include (1) completing only one energy block at GAST; (2) "conserving" the project; and (3) using GAST as a nuclear power research center but banning the housing of nuclear fuel at the station.
37. The Russian parliament's declaration of a five-year moratorium on new nuclear construction projects was signed by Yeltsin, June 28, 1990. (Associated Press Report [June 29, 1990].)
38. Unofficial copy of decision no. 21 available in author's personal archive.
39. Copy of letter available in author's personal archive.
40. Yurii Rogozhin, RSFSR Gosatomnadzor, interview by author (July 1993).

6. The National Enclaves: Tatarstan and Crimea

1. The bilateral treaty between the Russian Federation and the Republic of Tatarstan was signed on February 15, 1994. For more information on this treaty see Elizabeth Teague, "Russia and Tatarstan Sign Power-Sharing Treaty," *RFE/RL Research Report* 3, no. 14 (Apr. 8, 1994).

2. For more on this historical debate see Azade-Ayse Rorlich, *The Volga Tatars* (Stanford, C.A.: Hoover Institution Press, 1986), chapter 1.
3. See for example *Appeal to the Muslim Workers of Russia and the Soviet Far East,* issued by the Soviet People's Commissar (Nov. 20, 1917).
4. 1989 Census of the USSR, reported in A. Sheehy, "Russia's Republics: A Threat to Its Territorial Integrity?" *RFE/RL Research Report* 2, no. 20 (May 14, 1993).
5. Report on earlier coverage of the BVK debate published in *Vechernaya Kazan* (Oct. 6, 1987).
6. See for example the forum held at Dom Uchenyk in May 1987. Reported in *Vechernaya Kazan* (May 29, 1987).
7. *Vechernaya Kazan* (Oct. 6, 1987).
8. Longstanding support of local authorities noted in *Vechernaya Kazan* (Oct. 6, 1987).
9. Gosplan decision noted in *Vechernaya Kazan* (July 8, 1988).
10. For example, the campaign against a planned vinyl chloride factory in Kazan in 1988. Activists mobilized quickly on this issue and the government backed away from its construction plans almost immediately.
11. Plans for the Tatar AES were developed and approved in the 1970s. The station was planned to remedy the energy shortfall in the Tatar ASSR caused by the heavy and continuing industrialization of the republic. Moscow's decision to build the station was welcomed and approved by the Tatar Communist Party leadership as well as the Tatar Council of Ministers. *Kommunist Tatarii* 9 (1988).
12. *Sovetskaya Tatariya* (Apr. 8, 1989).
13. Activities of Kazan State University specialists are discussed in *Komsomolskaya pravda* (Aug. 18, 1989). Scientists from other institutions, including the Central Scientific-Research and Design Institute of Construction, the All-Union Scientific-Research Institute of the Geology of Non-Metallic Useful Metals, and the USSR Academy of Sciences, are also noted as important participants in this scientific debate. *Pravda* (May 10, 1989).
14. This perception of continued student apathy is supported by a group interview with members of the Kazan State University *druzhina* as well as by interviews with the environmental staff of the press office of Kazan State University and with the dean and chairman of the ecology department, Yurii Kotov. (Interviews conducted Mar. 1991.)
15. The environmental newsletter, alternately called *Greens of Tatariya* or *Magdi* first began to appear in September 1989. It was published on a monthly basis (rather sporadically) for approximately one year.
16. Note that this circle of activists that surrounded Garapov began to form in 1987 with the campaign against the biochemical BVK factory. At this time a small network of nonacademic activists was created which formed the basis for the ensuing anti-TAES crusade.
17. See for example the appeal against the Tatar AES published by Kotov and Konovalev in *Komsomolskaya pravda* (Aug. 18, 1989).

18. *Sovetskaya Rossiya* (April 26, 1989).
19. The march was reported in *Pravda* (May 10, 1989), and *Moscow News* 30 (July 23, 1989).
20. Such problems are discussed in an article by Albert Garapov, *Komsomolets Tatarii* (July 23, 1989).
21. Reported in *Komsomolskaya pravda* (Aug. 18, 1989).
22. Among the participants in these roundtables were Albert Garapov and Yurii Kotov, as well as the Chairman of the Tatar Gosplan and Deputy Chairman of the Tatar Council of Ministers, Yurii Voronin. Reported in *Komsomolskaya pravda* (Aug. 18, 1989).
23. *Komsomolskaya pravda* notes that over 200,000 people had signed petitions opposing the Tatar AES by the summer of 1989 (Aug. 18, 1989).
24. Reported in *Moscow News* 30 (July 23, 1989) and *Komsomolskaya pravda* (Aug. 18, 1989).
25. The march was reported in *Atmoda,* the newsletter of the Tatar People's Front (Nov. 20, 1989) and *Pravda* (Oct. 5, 1989).
26. Reported in *Sovetskaya Rossiya* (Oct. 22, 1989).
27. Reported in *Magdi* 4 (Nov. 1989).
28. Text of telegram reprinted in *Magdi* 4 (Nov. 1989).
29. Text of decree reprinted in *Sovetskaya Tatariya* (Nov. 14, 1989).
30. See for example survey conducted by the Tatar People's Front. *Atmoda* (Nov. 20, 1989).
31. This decision was taken on April 17, 1990. Resolution number 51-XII "On halting construction on the productive facility, Tatar AES." Unanimity of decision reported by Kolesnik, Chairman of the Ecology Committee of the Tatar Supreme Soviet, interview by author (Mar. 1991).
32. *Vechernaya Kazan* (Aug. 1, 1990).
33. Aleksei Kolesnik, interview by author (Mar. 1991).
34. A detailed report on these new informal groups is provided in *Kommunist Tatarii* 11 (1988).
35. Note that this group is alternately referred to in the media as the Tatar People's Front and the People's Front of Kazan. The formation of this initiative group was reported in *Kommunist Tatarii* 11 (1988).
36. See for example *Moscow News* 30 (July 23, 1989) and *Komsomolskaya pravda* (Aug. 18, 1989).
37. A particularly detailed discussion of anti-TAES activities is included in *Atmoda* (Nov. 20, 1989).
38. Because the Tatar Public Center is so often referred to by its Russian abbreviation (TOTS), I will use the Russian rather than the anglicized abbreviation in this discussion. TOTS stands for Tatarskii obshchestvennii tsentr.
39. Tatarstan adopted a declaration of sovereignty on August 30, 1990, which was confirmed and clarified in a national referendum held on March 22, 1992. The

referendum was passed by 61 percent of participating voters. See *Postfactum* (Mar. 22, 1992).

40. A bilateral treaty regulating economic and political relations between Russia and Tatarstan was signed on February 15, 1994. For more information on this treaty see Elizabeth Teague, "Russia and Tatarstan Sign Power-Sharing Treaty."

41. Note that while the Volga Tatars and Crimean Tatars can trace their heritage back to a common source from centuries ago, these two groups have evolved along different paths and now consider themselves to be entirely distinct ethnic groups. In this chapter, the term *Tatars* refers to the Volga Tatars and the Tatars in Crimea will always be referred to as the *Crimean Tatars.*

42. Zelinsky, director of public relations, Goskompriroda of Crimea, interview by author (May 22, 1991). KAES stands for Krymskaya atomnaya energichnaya stantsiya (or Crimean Atomic Power Station).

43. The group called itself "Problems of Natural Science."

44. The forum was advertised in the progressive paper, *Krymskii komsomolets.*

45. An interview with leading speakers from the April forum was published in *Krymskii komsomolets* (June 18, 1988).

46. See discussion of this forum in *Krymskii komsomolets* (May 14, 1988).

47. See for example the letter published in *Krymskaya pravda* (May 29, 1988), signed by writer Terekhov and scientist Svidzinsky, along with seventeen other writers and scientists.

48. Reported in *Krymskaya pravda* (May 21, 1988).

49. *Obkom* refers to the oblast committee of the Communist Party, the top party body in the oblast of Crimea.

50. This meeting was reported in *Krymskii komsomolets* (May 28, 1988) and *Krymskaya pravda* (June 18, 1988).

51. An interview with Melnikov revealed that his opposition to the station was based on the strength of public opinion on this issue. Two other delegates from Crimea, O. Mikhailets and V. Izmailov, were reportedly also in favor of cancellation, while a third, Crimean First Secretary A. Girenko, expressed sympathy with the opposition but doubted whether the region's energy needs could be met without the station. Interview reported in *Christian Science Monitor* (June 24, 1988).

52. See Boris Oleynik's speech to the conference, published in *V Sudbe Prirody—Nasha Sudba* (*In the Fate of Nature Lies Our Fate*) (Moscow: Khudozhestvennaya Literatura, 1990).

53. According to *Pravda* (Jan. 11, 1989) the official report was issued on November 28, 1988.

54. A letter signed by ten scientists was published in *Pravda* (Jan. 11, 1989). The scientists supported the commission's findings and called for a halt in construction of KAES. They also explicitly accused the Ministry of Atomic Energy of excluding specialist opinion.

In a letter published in *Robitnycha hazeta* (Jan. 8, 1989), vice president of the

Ukrainian Academy of Sciences, Lukinov, claims to be speaking on behalf of the Academy and notes that the Academy believes that the commission's findings indicate that construction on the KAES should be halted.
55. *Robitnycha hazeta* (May 20, 1989).
56. This letter was published in *Pravda Ukrainy* (Sept. 15, 1989).
57. Note that the Crimean party organization was supported by the Ukrainian Communist Party (UCP). The Central Committee of the UCP publicly voiced its support for canceling the KAES in an interview published in *Pravda Ukrainy* (May 24, 1989).
58. The decision was reported in *Izvestiya* (Oct. 2, 1989).
59. *Pravda* (Nov. 1, 1989).
60. The definition of a "Tatar state" is not always clear. For the most popular moderate organizations, such as the Organization of the Crimean Tatar National Movement, this term does not imply a monopoly on political rights by the Tatars. Rather, moderates call for a political system in which Tatars have the dominant political voice and the ability to veto any decisions taken by the Slavic population —despite the fact that the Crimean Tatars constitute only a small percentage of Crimea's population.
61. Crimean Tatars were estimated to make up 9.6 percent of Crimea's population in 1993. Andrew Wilson, *The Crimean Tatars* (London: International Alert, 1994).
62. Figure based on survey of Crimean population directed by Valerii Khmelko, Kiev State University, 1989.
63. For an excellent analysis of voting patterns in the 1994 Crimea elections see Andrew Wilson, "The Elections in Crimea," *RFE/RL Research Report* 3, no. 25 (June 24, 1994).

Conclusions

1. *ITAR-TASS* report of May 12, 1994, published in *Foreign Broadcast Information Service* (Central Eurasia) (May 13, 1994).
2. *Kreshchatik* (Kiev) (March 14, 1995): 2.
3. Note, for example, the picketing of the Russian Government building in Moscow by nuclear plant workers on April 5, 1994. Reported in *Rabochaya tribuna* (Apr. 13, 1995): 1.

SELECTED BIBLIOGRAPHY

Aleksandrov, A. "Scientific and Technological Progress in Nuclear Power Engineering." *World Marxist Review* 22, no. 6 (June 1979): 13–18.

Almond, Gabriel, and Sidney Verba. *The Civic Culture*. Princeton: Princeton University Press, 1963.

Apter, David, and Nagayo Sawa. *Against the State: Politics and Social Protest in Japan*. Cambridge: Harvard University Press, 1984.

Ash, Robert, and Mayer Zald. "Social Movement Organizations." *Social Forces* 44 (Mar. 1966): 327–341.

Barkan, Steven. "Strategic, Tactical and Organizational Dilemmas of the Protest Movement against Nuclear Power." *Social Problems* 27 (1979): 19–37.

Berry, Jeffrey. *Lobbying for the People*. Princeton, N.J.: Princeton University Press, 1977.

Breslauer, George. "In Defense of Sovietology." *Post-Soviet Affairs* 8, no. 3 (1992): 197–238.

Brubaker, Rogers. "Nationhood and the National Question in the Soviet Union and Post-Soviet Eurasia: An Institutional Analysis." *Theory and Society* (forthcoming, 1995).

Cohen, Jean. "Strategy or Identity: New Theoretical Paradigms and Contemporary Social Movements." *Social Research* 52, no. 4 (winter 1985): 663–716.

Comisso, Ellen. "Market Failures and Market Socialism: Economic Problems of the Transition." *Eastern European Politics and Societies* (fall 1988).

Conquest, Robert, ed. *The Last Empire: Nationality and the Soviet Future*. Stanford: Hoover Institution Press, 1986.

Darst, Robert. "Environmentalism in the USSR: The Opposition to the River Diversion Projects." *Soviet Economy* 4, no. 3 (July–Sept. 1988): 223–252.

———. "The Transnational Politics of Environmental Protection in the USSR and the Newly Independent States." Draft manuscript (1996).

Dawson, Jane. "The Soviet Scientific-Technical Intelligentsia: An Emerging Public?" In *Analyzing the Gorbachev Era,* edited by Gail Lapidus. Berkeley, Calif.: Berkeley-Stanford Program Publications, 1989, 87–116.

——. "Intellectuals and Anti-Nuclear Protest in the USSR." In *Beyond Sovietology: Essays in Politics and History,* edited by Susan Solomon. Armonk, N.Y.: M. E. Sharpe, 1993, 94–124.

——. "Anti-Nuclear Activism in the USSR and Its Successor States: A Surrogate for Nationalism?" *Environmental Politics* 4, no. 3 (1995): 441–466.

DeBardeleben, Joan. *The Environment and Marxism-Leninism: The Soviet and East German Experience.* Boulder and London: Westview Press, 1985.

DeBardeleben, Joan, and John Hannigan, eds. *Environmental Security and Quality after Communism.* Boulder, Colo.: Westview Press, 1995.

DeBoer, Connie, and Ineke Catsburg. "The Impact of Nuclear Accidents on Attitudes towards Nuclear Energy." *Public Opinion Quarterly* 52 (1988): 254–261.

Deudney, Daniel, and G. John Ikenberry. "The International Sources of Soviet Change." *International Security* 16, no. 3 (winter 1991–92): 74–118.

Dodd, C. K. *Industrial Decision-Making and High-Risk Technology: Siting Nuclear Power Facilities in the USSR.* Lanham, Md.: Rowman & Littlefield, 1994.

Dollezhal', N., and Yurii Koryakin. "Yadernaya energetika: Dostizheniya i problemi." *Kommunist* 14 (Sept. 1979): 19–28.

Douglas, Mary, and Aaron Wildavsky. *Risk and Culture: An Essay on the Selection of Technical and Environmental Dangers.* Berkeley: University of California Press, 1982.

Downs, Anthony. "Up and Down with Ecology—The Issue Attention Cycle." *The Public Interest* 28 (1972): 38–50.

Duka, A., N. Kornev, V. Voronkov, and E. Zdravomyslova. "Roundtable on Russian Sociology: The Protest Cycle of Perestroika." *International Sociology* 10, no. 1 (1995).

Dunlop, John. "Russia: Confronting a Loss of Empire." In *Nations and Politics in the Soviet Successor States,* edited by Ian Bremmer and Ray Taras. Cambridge: Cambridge University Press, 1993.

Ebbin, S., and Raphael Kasper. *Citizen Groups and the Nuclear Power Controversy: Uses of Scientific and Technological Information.* Cambridge: MIT Press, 1974.

Etzioni, Amitai. *The Active Society.* New York: Free Press of Glencoe, 1968.

Feshbach, Murray. *Ecocide in the USSR: Health and Nature under Seige.* New York: Basic Books, 1992.

——. *Ecological Disaster.* New York: Twentieth-Century Fund Press, 1995.

Freeman, John. "Resource Mobilization and Strategy." In *The Dynamics of Social Movements,* edited by John McCarthy and Mayer Zald. Cambridge: Winthrop, 1979.

Freudenburg, W., and Eugene Rosa, eds. *Public Reaction to Nuclear Power: Are There Critical Masses?* Boulder, Colo.: Westview Press, 1984.

Fuller, Elizabeth. "Armenian Authorities Appear to Yield to Ecological Lobby." *RL Research Bulletin* 130 (1987).

Galeyeva, A., and M. Kurok, eds. *Ob okhrane okruzhayushchei sredy. Sbornik dokumentov partii i pravitel'stva, 1917–1985 gg.* Moscow: Politizdat, 1986.

Gamson, William. *The Strategy of Social Protest.* Homewood, Ill.: Dorsey, 1975.

Gamson, William, and Bruce Fireman. "Utilitarian Logic in the Resource Mobilization Perspective." In *The Dynamics of Social Movements,* edited by John McCarthy and Mayer Zald. Cambridge: Winthrop, 1979.

Gellner, Ernest. *Nations and Nationalism.* Ithaca, N.Y.: Cornell University Press, 1983.

Gerhards, Jurgen, and Dieter Rucht. "Mesomobilization: Organizing and Framing in Two Protest Campaigns in West Germany." *American Journal of Sociology* 98 (1992): 555–596.

Gerlach, Luther, and Virginia Hine. *People, Power, Change.* New York: Bobbs-Merrill, 1970.

Girnius, S. "Continued Controversy over Third Reactor at Ignalina Atomic Power Plant." *RL Research Bulletin* (Aug. 4, 1988).

———. "Protest at Ignalina Atomic Power Station." *RL Research Bulletin* (Sept. 19, 1988).

Goble, Paul. "Russia and Its Neighbors." *Foreign Policy* 90 (spring 1993): 79–88.

Goldman, Marshall. *The Spoils of Progress: Environmental Pollution in the Soviet Union.* Cambridge, Mass.: MIT Press, 1972.

Gould, Peter. *Fire in the Rain: The Democratic Consequences of Chernobyl.* Cambridge: Polity Press, 1990.

Green, E. *Ecology and Perestroika: Environmental Protection in the Soviet Union.* Washington, D.C.: American Committee on U.S.-Soviet Relations, 1990.

Gurr, Ted. *Why Men Rebel.* Princeton, N.J.: Princeton University Press, 1970.

Gustafson, Thane. *Reform in Soviet Politics.* Cambridge: Cambridge University Press, 1981.

———. *Crisis amid Plenty: The Politics of Soviet Energy under Brezhnev and Gorbachev.* Princeton, N.J.: Princeton University Press, 1989.

Habermas, Jurgen. "New Social Movements." *Telos* 48 (fall 1981): 33–37.

———. *Theory of Communicative Action,* vol. 1. Boston: Beacon Press, 1984.

Hajda, Lubomyr, and Mark Beissinger, eds. *The Nationalities Factor in Soviet Politics and Society.* Boulder, Colo.: Westview Press, 1990.

Halverson, Thomas. "Ticking Time Bombs: East Bloc Reactors." *The Bulletin of the Atomic Scientists* 49, no. 6 (July/Aug. 1993): 43–48.

Hannigan, Joan. "Alain Touraine, Manuel Castells and Social Movement Theory." *Sociological Quarterly* 26 (1985): 435–454.

Hewett, Ed. *Reforming the Soviet Economy: Equality versus Efficiency.* Washington, D.C.: The Brookings Institution, 1988.

Hewett, Ed, and Victor Winston, eds. *Milestones in Glasnost and Perestroyka: The Economy.* Washington, D.C.: The Brookings Institution, 1991.

Hirschman, Albert. *Exit, Voice and Loyalty.* Cambridge: Harvard University Press, 1970.
Hobsbawn, Eric. *Nations and Nationalism since 1980.* Cambridge: Cambridge University Press, 1990.
Holdar, Sven. "Torn between East and West: The Regional Factor in Ukrainian Politics." *Post-Soviet Geography* 36, no. 2 (1995).
Horowitz, David. *Ethnic Groups in Conflict.* Berkeley: University of California Press, 1985.
Hough, Jerry. *The Soviet Union and Social Science Theory.* Cambridge: Harvard University Press, 1977.
IAEA. *Summary Report on the Post-Accident Review Meeting on the Chernobyl Accident.* Report by the International Nuclear Safety Advisory Group. Safety Series no. 75-INSAG-1. Vienna: International Atomic Energy Agency, 1986.
——. *Operational Safety of Nuclear Installations: Rovenskaya Nuclear Power Plant, 5-22 December 1988.* Report to the Government of the USSR. IAEA-NENS/OSART/89/20. Vienna: International Atomic Energy Agency, 1989.
——. *The International Chernobyl Project: An Overview.* Report by an International Advisory Committee. Vienna: International Atomic Energy Agency, 1991.
——. *The International Chernobyl Project: Proceedings of an International Conference.* Vienna: International Atomic Energy Agency, 1991.
——. *The International Chernobyl Project: Surface Contamination Maps.* Vienna: International Atomic Energy Agency, 1991.
——. *Nuclear Power Reactors in the World.* April 1992 Edition, Reference Data Series No. 2. Vienna: International Atomic Energy Agency, 1992.
Idzelis, A. "The Socioeconomic and Environmental Impact of the Ignalina Nuclear Power Station." *Journal of Baltic Studies* XIV, no. 3 (fall 1983): 247–254.
Inglehart, Ronald. *The Silent Revolution: Changing Values and Political Styles among Western Publics.* Princeton, N.J.: Princeton University Press, 1977.
——. "The Fear of Living Dangerously: Public Attitudes toward Nuclear Power." *Public Opinion* 6 (Feb./Mar. 1984): 41–44.
Jancar, B. *Environmental Management in the Soviet Union and Yugoslavia.* Durham, N.C.: Duke University Press, 1987.
Janos, Andrew. "Group Politics in Communist Society: A Second Look at the Pluralist Model." In *Authoritarian Politics in Modern Society,* edited by Samuel Huntington and Clement Moore. New York: Basic Books, 1970.
——. "Interest Groups and the Structure of Power: Critique and Comparisons." *Studies in Comparative Communism* XII, no. 1 (spring 1979): 6–20.
Jenkins, J. Craig. "The Transformation of a Constituency into a Movement." In *The Social Movements of the 1960s and 1970s,* edited by John Freeman. New York: Longmans, 1982.
——. "Resource Mobilization Theory and the Study of Social Movements." *Annual Review of Sociology* 9 (1983): 527–553.

Joppke, Christian. *Mobilizing against Nuclear Energy: A Comparison of Germany and the United States*. Berkeley: University of California Press, 1993.

Jowitt, Kenneth. "An Organizational Approach to the Study of Political Culture in Marxist-Leninist Systems." *American Political Science Review* LXVIII, no. 3 (Sept. 1974): 1171–1191.

——. *The Leninist Response to National Dependency*. Berkeley, Calif.: Institute for International Studies, 1978.

——. "Soviet Neotraditionalism: The Political Corruption of a Leninist Regime." *Soviet Studies* (July 1983): 275–297.

Karatnycky, A. "The Ukrainian Factor." *Foreign Affairs* 71, no. 3 (summer 1992): 90–107.

Kassof, Allen. "The Administered Society: Totalitarianism without Terror." *World Politics* (July 1964): 558–575.

Katsman, D. *Soviet Nuclear Power Plants: Reactor Types, Water and Chemical Control Systems, Turbines*. Falls Church, Va.: Delphic Associates, 1986.

Keane, John. *Civil Society and the State*. London: Verso, 1988.

Kelley, Donald. "Environmental Policy-Making in the USSR: The Role of Industrial and Environmental Interest Groups." *Soviet Studies* 28 (Oct. 1976): 570–589.

Kitschelt, Herbert. "Political Opportunity Structures and Political Protest: Anti-Nuclear Movements in Four Democracies." *British Journal of Political Science* 16 (1986).

Klandermans, Bert, and Sidney Tarrow. "Mobilization and Social Movements: Synthesizing European and American Approaches." *International Social Movement Research* 1 (1988).

Kneen, Peter. *Soviet Scientists and the State*. Albany, N.Y.: SUNY Press, 1984.

Kolsto, Pal. "The New Russian Diaspora: Minority Protection in the Soviet Successor States." *Journal of Peace Research* 30, no. 2 (1993).

Komarov, Boris. *The Destruction of Nature in the Soviet Union*. White Plains, N.Y.: M. E. Sharpe, 1980.

Kornhauser, William. *The Politics of Mass Society*. New York: Free Press of Glencoe, 1959.

Kramer, J. "Chernobyl and Eastern Europe." *Problems of Communism* 35, no. 6 (Nov.–Dec. 1986).

——. "The Nuclear Power Debate in Eastern Europe." *RFE/RL Research Report* 1, no. 35 (Sept. 4, 1992): 59–65.

Krawchenko, Bohdan. "Ukraine: The Politics of Independence." In *Nations and Politics in the Soviet Successor States*, edited by Ian Bremmer and Ray Taras. Cambridge: Cambridge University Press, 1993.

Krickus, Richard. "Lithuania: Nationalism in the Modern Era." In *Nations and Politics in the Soviet Successor States*, edited by Ian Bremmer and Ray Taras. Cambridge: Cambridge University Press, 1993.

Laitin, David. "The National Uprisings in the Soviet Union." *World Politics* 44, no. 1 (Oct. 1991): 139–177.

Lapidus, Gail. “Gorbachev and the Reform of the Soviet System.” *Daedalus* 116, no. 2 (spring 1987).
Lazzerini, Edward. “Crimean Tatars.” In *The Nationalities Question in the Post-Soviet States*, edited by Graham Smith. London: Longman Press, 1996, 412–435.
Lewin, Moshe. *The Gorbachev Phenomenon*. Berkeley: University of California Press, 1988.
Lowenhardt, John. *Decision Making in Soviet Politics*. New York: St. Martin’s Press, 1981.
Lowenthal, Richard. “Development versus Utopia in Communist Policy.” In *Change in Communist Systems,* edited by Chalmers Johnson. Stanford, Calif.: Stanford University Press, 1970, 33–116.
Lubrano, Linda, and Susan Solomon, eds. *The Social Context of Soviet Science*. Folkestone: Wm Dawson, 1980.
Mackay, L., and M. Thompson, eds. *Something in the Wind: Politics after Chernobyl*. London: Pluto, 1988.
Marples, David. *Chernobyl and Nuclear Power in the USSR*. New York: St. Martin’s Press, 1986.
——. *The Social Impact of the Chernobyl Disaster*. New York: St. Martin’s Press, 1988.
——. *Ukraine under Perestroika*. New York: St. Martin’s Press, 1991.
Marwell, Gerald, and Ames. “Experiments on the Provision of Public Goods, I & II.” *American Journal of Sociology* 84 & 85 (1979 & 1980).
Marx, Gary, and J. Wood. “Strands of Theory and Research in Collective Behavior.” *Annual Review of Sociology* 1 (1978): 363–428.
McAdam, Doug, John McCarthy, and Mayer Zald. “Social Movements.” In *Handbook of Sociology,* edited by Neil Smelser. Newbury Park, Calif.: Sage Publications, 1988.
McCarthy, John, and Mayer Zald. *The Trend of Social Movements*. Morristown, N.J.: General Learning, 1973.
——. “Resource Mobilization and Social Movements.” *American Journal of Sociology* 82 (1977).
McKean, Margaret. *Environmental Protest and Citizen Politics in Japan*. Berkeley: University of California Press, 1981.
Medvedev, Grigorii. *The Truth about Chernobyl*. New York: Basic Books, 1991.
Medvedev, Zhores. *Nuclear Disaster in the Urals*. London: Angus and Robertson, 1979.
——. “The Soviet Nuclear Energy Programme: The Road to Chernobyl.” In *Something in the Wind: Politics after Chernobyl*, edited by L. Mackay and M. Thompson. London: Pluto, 1988.
——. *The Legacy of Chernobyl*. New York and London: W. W. Norton, 1990.
Melucci, Albert. “The Symbolic Challenge of Contemporary Movements.” *Social Research* 52, no. 4 (winter 1985): 789–816.

———. "Getting Involved: Identity and Mobilization in Social Movements." In *International Social Movement Research,* edited by Bert Klandermans, Hanspeter Kriesi, and Sidney Tarrow. Greenwich, Conn.: JAI Press, 1988.

Meyer, M. "USSR Incorporated." *Slavic Review* XX (1961): 369–376.

Millard, Frances. "The Polish Response to Chernobyl." *Journal of Communist Studies* 4 (1988): 27–53.

Misiunas, R. "The Baltic Republics." In *The Nationalities Factor in Soviet Politics and Society,* edited by Lubomyr Hajda and Mark Beissinger. Boulder, Colo.: Westview Press, 1990.

Moberg, A. *Before and After Chernobyl: Nuclear Power in Crisis.* Amsterdam: WISE, 1986.

Moore, Barrington. *Terror and Progress — USSR.* Cambridge: Harvard University Press, 1954.

Morone, Joseph, and Edward Woodhouse. *The Demise of Nuclear Energy? Lessons for the Democratic Control of Technology.* New Haven, Conn.: Yale University Press, 1989.

Morris, Aldon, and Cedric Herring. "Theory and Research in Social Movements: Critical Review." In *Political Behavior Annual,* edited by S. Long. Boulder, Colo.: Westview Press, 1984.

Morris, Aldon, and Carol Mueller. *Frontiers in Social Movement Theory.* New Haven: Yale University Press, 1992.

Motyl, Alexander. *Dilemmas of Independence: Ukraine after Totalitarianism.* New York: Council on Foreign Relations Press, 1993.

———, ed. *Thinking Theoretically about Soviet Nationalities.* New York: Columbia University Press, 1992.

Nahaylo, Bohdan. "Baltic Echoes in Ukraine." In *Toward Independence: The Baltic Popular Movements,* edited by J. Trapans. Boulder, Colo.: Westview Press, 1991, 109–122.

Nelkin, Dorothy and Michael Pollak. *The Atom Besieged: Extraparliamentary Dissent in France and Germany.* Cambridge: MIT Press, 1981.

Oberschall, Anthony. *Social Conflict and Social Movements.* Englewood Cliffs, N.J.: Prentice Hall, 1973.

O'Donnell, Guillermo, Philippe Schmitter, and Laurence Whitehead, eds. *Transitions from Authoritarian Rule.* Baltimore, Md.: Johns Hopkins University Press, 1986.

Offe, Claus. "New Social Movements: Challenging the Boundaries of Institutional Politics." *Social Research* 52 (1985).

Olson, Mancur. *The Logic of Collective Action.* New York: Schocker, 1968.

Perrow, Charles. *Normal Accidents: Living with High Risk Technologies.* New York: Basic Books, 1984.

Peterson, D. J. *Troubled Lands: The Legacy of Soviet Environmental Destruction.* Boulder, Colo.: Westview Press, 1993.

Piven, Frances, and Richard Cloward. *Poor People's Movements.* New York: Pantheon, 1977.

Pizzorno, Alessandro. "Political Exchange and Collective Identity in Industrial Conflict." In *The Resurgence of Class Conflict in Western Europe since 1968,* edited by Crouch and Pizzorno. London: Macmillan, 1978, 277–298.

Potter, William. "Soviet Decision Making for Chernobyl: An Analysis of System Performance and Policy Change." Report to the National Council for Soviet and East European Research (1989).

———. "The Impact of Chernobyl on Nuclear Power Safety in the Soviet Union." *Studies in Comparative Communism* 24, no. 2 (June 1991): 191–210.

Pryde, Philip. *Environmental Management in the Soviet Union.* New York: Cambridge University Press, 1991.

———, ed. *Environmental Resources and Constraints in the Former Soviet Republics.* Boulder, Colo.: Westview Press, 1995.

Rigby, T. "Organizational, Traditional, and Market Societies." *World Politics* (July 1964): 539–557.

Roeder, Philip. "Soviet Federalism and Ethnic Mobilization." *World Politics* 43, no. 2 (Jan. 1991): 196–232.

Rohrschneider, Robert. "Citizens' Attitudes toward Environmental Issues: Selfish or Selfless? *Comparative Political Studies* 21, no. 3 (Oct. 1988): 347–367.

Ruble, Blair. "The Soviet Union's Quiet Revolution." In *Can Gorbachev's Reforms Succeed?* edited by George Breslauer. Berkeley, Calif.: Berkeley-Stanford Publication, 1990.

Rudig, Wolfgang. *Anti-Nuclear Movements: A World Survey of Opposition to Nuclear Power.* Essex, U.K.: Longman Group, 1990.

Scanlon, John. "Reform and Civil Society in the USSR." *Problems of Communism* (Apr. 1988): 41–46.

Schoenfeld, A., R. Meier, and R. Griffin. "Constructing a Social Problem: The Press and the Environment." *Social Problems* 27 (1979).

Selznick, Philip. *The Organizational Weapon: A Study of Bolshevik Strategy and Tactics.* New York: McGraw-Hill, 1952.

Senn, Albert. *Lithuania Awakening.* Berkeley: University of California Press, 1990.

Shaw, Dennis, and M. Bradshaw. "Problems of Ukrainian Independence." *Post-Soviet Geography* 33, no. 1 (Jan. 1992).

Shcherbak, Yurii. *Chernobyl: A Documentary Story.* London: Macmillan Press, 1989.

Silver, Brian. "Political Beliefs of the Soviet Citizen." In *Politics, Work, and Daily Life in the USSR,* edited by James Millar. Cambridge: Cambridge University Press, 1987, 100–141.

Skilling, Gordon. "Interest Groups and Communist Politics." *World Politics* XVIII, no. 3 (Apr. 1966): 435–51.

Skilling, Gordon, and Franklyn Griffiths. *Interest Groups in Soviet Politics.* Princeton, N.J.: Princeton University Press, 1971.

Skocpol, Theda. *States and Social Revolutions*. Cambridge: Cambridge University Press, 1979.
Smelser, Neil. *Theory of Collective Behavior*. New York: Free Press of Glencoe, 1963.
Smith, Graham, ed. *The Nationalities Question in the Post-Soviet States*. London: Longman Press, 1996.
Smith, Kathleen. *Remembering Stalin's Victims*. Ithaca, N.Y.: Cornell University Press, 1996.
Snyder, Jack. "Nationalism and the Crisis of the Post-Soviet State." *Survival* 35, no. 1 (Spring 1993).
Solchanyk, Roman. "Ukraine, the (Former) Center, Russia, and 'Russia.'" *Studies in Comparative Communism* 25, no. 1 (Mar. 1992): 31–45.
——. *Ukraine: From Chernobyl to Sovereignty*. New York: St. Martin's Press, 1992.
——. "The Politics of State Building: Centre-Periphery Relations in Post-Soviet Ukraine." *Europe-Asia Studies* 46, no. 1 (1994): 47–68.
Solomon, Peter. *Soviet Criminologists and Criminal Policy*. New York: Columbia University Press, 1978.
Starr, Frederick. "Soviet Union: A Civil Society." *Foreign Policy* 70 (spring 1988): 26–41.
Subtelny, Ornest. *Ukraine: A History*. Toronto: University of Toronto Press, 1989.
Szporluk, Roman. "The Ukraine and Russia." In *The Last Empire: Nationality and the Soviet Future,* edited by Robert Conquest. Stanford, Calif.: Hoover Institution Press, 1986, 151–182.
——. "The Imperial Legacy and the Soviet Nationalities Problem." In *The Nationalities Factor in Soviet Politics and Society,* edited by L. Hajda and M. Beissinger. Boulder, Colo.: Westview Press, 1990.
Tarrow, Sidney. "Mentalities, Political Cultures, and Collective Action Frames: Constructing Meanings through Action." In *Frontiers in Social Movement Theory,* edited by Aldon Morris and Carol Mueller. New Haven, Conn.: Yale University Press, 1992.
Thornton, Judith. "Soviet Electric Power after Chernobyl: Economic Consequences and Options." *Soviet Economy* 2, no. 2 (1986): 131–179.
Tilly, Charles. *From Mobilization to Revolution*. Reading, Mass.: Addison-Wesley, 1978.
——. "Models and Realities of Popular Collective Action." *Social Research* 52, no. 4 (winter 1985): 715–747.
Touraine, Alain. *The Voice and the Eye*. New York: Cambridge University Press, 1981.
——. *Anti-Nuclear Protest*. Cambridge: Cambridge University Press, 1983.
——. "An Introduction to the Study of Social Movements." *Social Research* 52, no. 4 (winter 1985).

Turner, R. "Political Opposition to Nuclear Power: An Overview." *Political Quarterly* 57 (1986).

Turner, Ralph, and Lewis Killian. *Collective Behavior.* Englewood Cliffs, N.J.: Prentice Hall, 1972, 438–443.

Urban, Michael, ed. *Ideology and System Change in the USSR and Eastern Europe.* New York: St. Martin's Press, 1992.

———. "The Politics of Identity in Russia's Postcommunist Transition: The Nation against Itself." *Slavic Review* 35, no. 3 (fall 1994): 733–765.

Vardys, V. S. "Sajudis: National Revolution in Lithuania." In *Toward Independence: The Baltic Popular Movements,* edited by Jan Trapans. Boulder, Colo.: Westview Press, 1991, 11–23.

Varley, J. "Three Years after Chernobyl: Soviet Public Opposition Grows." *Nuclear Engineering International* (Apr. 1989).

Voronitsyn, S. "The Chernobyl Disaster and the Myth of the Safety of Nuclear Power Stations." *Report on the USSR* 202 (1986).

Walder, Andrew. *Communist Neotraditionalism: Work and Authority in Chinese Industry.* Berkeley: University of California Press, 1986.

Weber, Max. "The Types of Legitimate Domination." In *Economy and Society,* 212–301. Berkeley: University of California Press, 1978.

Weiner, Doug. *Models of Nature: Ecology, Conservation, and Cultural Revolution in Soviet Russia.* Bloomington: Indiana University Press, 1988.

Wilson, Andrew. *The Crimean Tatars.* London: International Alert, 1993.

Wilson, William J. *Power, Racism and Privilege.* New York: Free Press of Glencoe, 1973.

Wilson, James Q. *Political Organizations.* New York: Basic, 1973.

Yanitsky, Oleg. *Russian Environmentalism: Leading Figures, Facts, Opinions.* Moscow: Mezhdunarodniye Otnosheniya, 1993.

Zaslavskaya, Tatyana. "The Novosibirsk Report." *Survey* 28, no. 1 (spring 1984): 88–108.

Zeigler, Charles. *Environmental Policy in the USSR.* Amherst: University of Massachusetts Press, 1987.

INDEX

About the Author

Jane I. Dawson is Assistant Professor of Political Science at the University of Oregon.

Library of Congress Cataloging-in-Publication Data

Dawson, Jane I.

Eco-nationalism : anti-nuclear activism and national identity in Russia, Lithuania, and Ukraine / Jane I. Dawson.

p. cm.

Includes bibliographical references (p.) and index.

ISBN 0-8223-1831-8 (alk. paper). —

ISBN 0-8223-1837-7 (pbk. : alk. paper)

1. Nuclear industry — Ukraine. 2. Nuclear industry — Russia (Federation). 3. Nuclear industry — Lithuania. 4. Antinuclear movement — Ukraine. 5. Antinuclear movement — Russia (Federation) 6. Antinuclear movement — Lithuania. 7. Urkraine — History — Autonomy and independence movements. 8. Russia (Federation) — History — Autonomy and independence movements. 9. Lithuania — History — Autonomy and independence movements. I. Title.

HD9698.U382D39 1996

333.792′4′0947 — dc20

96-14630

CIP